# MECHANICAL VIBRATIONS AND CONDITION MONITORING

# MECHANICAL VIBRATIONS AND CONDITION MONITORING


JUAN CARLOS A. JAUREGUI CORREA

*Autonomous University of Queretaro*
*Queretaro, Mexico*

ALEJANDRO A. LOZANO GUZMAN

*National Polytechnic Institute*
*Queretaro, Mexico*



ELSEVIER

ACADEMIC PRESS

An imprint of Elsevier

Academic Press is an imprint of Elsevier
125 London Wall, London EC2Y 5AS, United Kingdom
525 B Street, Suite 1650, San Diego, CA 92101, United States
50 Hampshire Street, 5th Floor, Cambridge, MA 02139, United States
The Boulevard, Langford Lane, Kidlington, Oxford OX5 1GB, United Kingdom



**Library of Congress Cataloging-in-Publication Data**
A catalog record for this book is available from the Library of Congress

**British Library Cataloguing-in-Publication Data**
A catalogue record for this book is available from the British Library

ISBN 978-0-12-819796-7

For information on all Academic Press publications
visit our website at https://www.elsevier.com/books-and-journals

*Publisher:* Matthew Deans
*Acquisitions Editor:* Glyn Jones
*Editorial Project Manager:* Rafael G. Trombaco
*Production Project Manager:* Poulouse Joseph
*Cover Designer:* Harris Greg

Typeset by SPi Global, India

# Contents

# Introduction

Conditioning monitoring is an evolution of predictive maintenance or pro-active maintenance. The origin is difficult to define, but predictive maintenance has made enormous progress over the last few decades. Nowadays, it has been addressed as one of the most innovative solutions for anticipating failures in machinery and is being used by a wide variety of industrial sectors. Prediction maintenance could be applied to a large industrial sector when the cost of vibration sensors was competitive. This advantage reduced the cost of failures in comparison with the investment in the measurement equipment and analysis system. In the beginning, the systems were rudimentary and required specialized personnel to collect data and make the analysis. The convergence of high-precision accelerometers and the ability to process the Fourier transform with the FFT algorithm allowed the development of rapid tools that can make a diagnosis on the actual condition of machines. Previously, vibration sensors were applied in just a few types of equipment due to their cost and the need for specialized personnel. These first concepts were complemented with other emerging technologies such as ultrasonic, thermography, acoustic sensors, and directional microphones.

The first application of predictive maintenance was made by the Royal Air Force in the United Kingdom. It was found that the rate of failure increased after the repair or inspection of machines, even following the maintenance plans. This phenomenon was named the "Waddington Effect," which led to condition monitoring. It was decided to adjust the maintenance programs and align them to the physical condition and frequency of use to reduce the Waddington effect. The process required the analysis of much data, but the launch of this program reduced the number of failures. Conditioning monitoring systems evaluate the vibration data and determine the condition of the machine based on the analysis of amplitude and frequency. The original signal has raw data that have to be treated to produce a reference baseline. Sampling the evolution of the data during operation indicates the condition of the machine and, in the case of a failure, the data will present significant changes. Conditioning monitoring systems have increased the reliability of machinery because they include new sensors while also using fast processing hardware and better algorithms for the signal process. The application of artificial intelligent programs in conditioning monitoring systems has increased the reliability of modern machinery, allowing more extended periods between maintenance. These complex systems can anticipate failures in most of the components that constitute modern machines. The analysis data is also linked to purchasing programs and the supply chain, enabling the reduction of spare part inventories.

Conditioning monitoring systems depend on the diagnosis of machinery, equipment, and machine trains. The principal source of data for the diagnosis comes from measuring mechanical vibrations. Traditionally, the study of the mechanical vibrations has been considered tedious work without an immediate application in engineering practice in the industry, but that could not be further from reality. The vibratory phenomenon is, practically, present in all machines and mechanical systems of any industrial facility or structure. The force that excites any structure comes from the movement of mechanical components. The variations in the homogeneity of the materials are among the causes of vibrations. Other sources of vibrations are the imperfections in the machining process, the manufacturing tolerances, and the clearance for assembly of the machines. A machine will vibrate if it operates in an overloaded state, if it is working below its design parameters, if it lacks maintenance, or if it has excessive wear. The intention of measuring vibrations is to have enough information to analyze the conditions of a machine train and to keep the vibration levels of the machines within acceptable values. Also, with proper interpretation, these vibrations can provide a great deal of information regarding the operating conditions of the equipment and mechanical systems.

The development of electronics, computers, and software provides great storage capacity and information processing, which make possible the application of vibration analysis in an industrial environment. However, it is necessary to have an in-depth knowledge of the operating principles of the measuring devices. The measuring systems generate large amounts of data that must be analyzed, taking into account the theoretical background (theory of vibrations) and the analysis tools. The combination of these elements potentializes the application of Conditioning Monitoring Systems. Otherwise, one can easily become a "black box" user without being able to interpret the causes and effects that mechanical vibrations have in the different mechanical systems.

This book has two main objectives: To provide a maintenance engineer with the necessary tools to make proper use of his/her measuring equipment, and to introduce both engineering students and recently qualified engineers in the practical implementation of vibration analysis in the predictive maintenance of machinery. The mathematical developments highlight the principal concepts for understanding the theory, emphasizing the ideas of the vibration theory as well as their importance in the practice of predictive maintenance.

In particular, the book addresses the technicians and engineers who need to apply their knowledge in a practical way to conditioning monitor a machine train. Likewise, it will also be useful for actual technicians to support their observations and decisions theoretically concerning machinery and equipment maintenance.

## General considerations

The idea that the maintenance of the machinery is fundamental to increase the productivity in an industrial plant is a concept that, finally, is being well taken among engineers. This is surprising considering the number of industrial executives who have ignored the impact that maintenance has in the quality and quantity of the products they manufacture. The results are evident because the maintenance practices are increasing and improving, along with the consequent reduction of costs and incidences in production.

Three maintenance administration techniques can be identified: The failure of the equipment or corrective maintenance, the scheduled regular intervals (preventive maintenance), and associating maintenance with the analysis of the performance of the equipment (predictive maintenance).

The first of the aforementioned techniques is simple and direct: When a machine fails, it has to be fixed. In a plant where this maintenance technique is put into practice, money is spent only on spare parts when the machine stops operating. However, in general it is the most expensive technique for the costs involved due to the lack of production and the negative effects this failure causes in the operation of the rest of the plant. In fact, a total lack of maintenance is unusual in most companies, considering that some basic lubrication and minor adjustment activities are made in the machines. However, this minimal maintenance also requires having the capacity to fix all machines of the plant while also getting spare parts and technical assistance within a minimum emergency time, which increases the costs of both the technical advice and spare parts.

A more rational approach to plant maintenance is constituted thanks to so-called preventive maintenance, which bases its methodology on the statistical behavior of the different pieces of equipment of industrial processes or in a given mechanical system. There is a lot of information that shows that mechanical systems, especially those in industrial plant machines, are more susceptible to fail at the beginning of their lifespan due to installation errors and inherent mismatches to their manufacture. This period is known as settlement. Immediately, the failure rate decreases considerably.

## Acknowledgments

Achieving knowledge and spreading it is perhaps one of the most significant aspects of our society. The social impact of exchanging ideas and points of view allowed us to create this book. We appreciate the effort and support provided by diverse professionals, most of whom are researchers and specialists who helped enrich our analysis with their contributions and recommendations. We must also recognize the institutional support given to us, for welcoming and stimulating our research and scientific proposal, to the Universidad Autónoma de Querétaro and the Instituto Politécnico Nacional.

In particular, we express our gratitude to Luis Huerta Ishikawa for his thoughtful comments regarding the material of the book. Likewise, to Ricardo Santamaria Holek for his wise initiative to integrate and publish the present work. Also, we want to thank Frank Beker Rubio for his effort in reviewing the quality of the translation. Finally, we thank Rodolfo Vega Hernandez for his support in translating the original material to English.

**Juan Carlos A. Jauregui Correa**
**Alejandro A. Lozano Guzman**

CHAPTER ONE

# Fundamentals of mechanical vibrations

## General considerations

The basis for most conditioning monitoring systems is the analysis of vibration signals. Thus, a fundamental part to understand the causes and effects of vibrations is the study of oscillatory movements and the interactions among the different components of a machine or a set of machines. These oscillatory movements, in the context of design and machinery analysis, are known as mechanical vibrations or, more generically, vibrations. The essence of conditioning monitoring is the analysis of the relationship between the input signal (the source of vibration) and the output response (the output signal), and the evolution of the dynamic behavior of the machine. In most cases, the machinery can be considered as a lineal system, although there are some particular cases that will be analyzed in other chapters, and the output signal will be a linear response of the excitation forces. Even though a machine is a complex system composed of a large number of mechanical elements, its dynamic response can be represented as a simple lumped-mass system.

The presented approach highlights the basic concepts of mechanical vibrations, leaving the mathematical developments for further consultation. There are a great number of excellent textbooks devoted to the study of mechanical vibrations, presenting solution methods for movement equations of the different cases; from these textbooks, only the essential concepts are summarized. The material presented in this book is organized to understand the basis for the application of a conditioning monitoring system and the interpretation for the diagnosis of machinery.

At the same time, a link is established between the merely practical and the empirical study of vibrations and the formal study of the basic concepts of the mechanical vibrations theory. The concepts presented in this chapter will allow the maintenance engineer to analyze and predict causes and effects in the operation of the machines.

*Mechanical Vibrations and Condition Monitoring*
https://doi.org/10.1016/B978-0-12-819796-7.00001-9

Among the concepts presented in this chapter, the definition of mechanical vibrations will be divided into deterministic and random vibrations. The deterministic vibrations represent regular movements whose waveform can be known in time, whereas random vibrations do not repeat regularly. However, this chapter presents the fundamentals of the study of the mechanical vibration, considering that they are deterministic, suggesting that the movement is harmonic, that is, that it regularly repeats in time.

Random vibrations are usually found in the vibration analysis for conditioning monitoring. However, through spectral analysis techniques, random signals are studied by applying the defined concepts for the harmonic vibrations.

In order to analyze the vibratory phenomena, it is necessary to represent the movements so that the characteristics of the vibrations can be identified as either deterministic or random. The simplest representation of a machine is made by idealizing its vibratory movements as a mass in motion supported by a spring and a viscous damper (Fig. 1.1). The mass is considered a single particle and the spring is considered an element that obeys Hook's law, where $k$ is the stiffness constant of the spring, $m$ is the mass of the particle, and the viscous constant $c$ is the damping of this movement. The force generated by oscillating this spring is given by $F_r = kx$, where $x$ is the displacement that the mass $m$ suffers. The force generated by this damper is given by $F_a = cv$, where $c$ is the damping coefficient and $v$ is the speed of the mass motion $m$. The external forces applied to the mechanical system formed by the mass, the spring, and the damper are represented by $F(t)$.

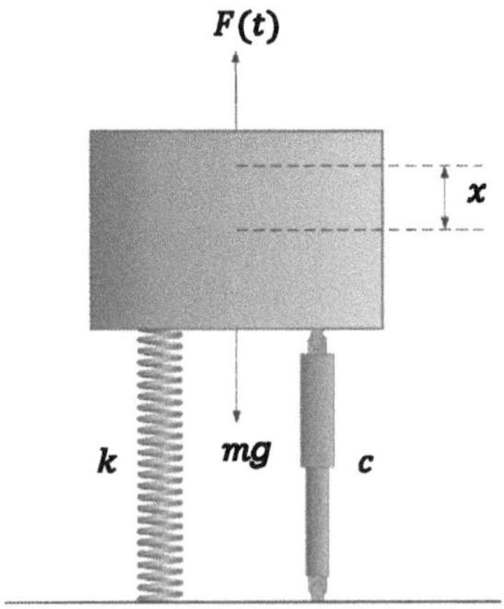


**Fig. 1.1** Conceptual model of the vibratory system.

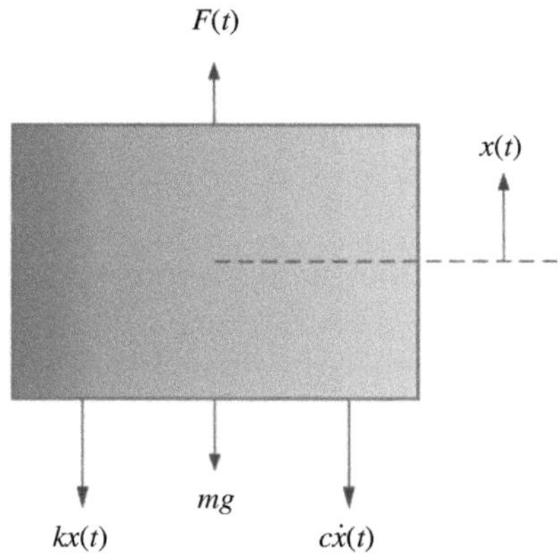


**Fig. 1.2** Representation of the acting forces on a vibratory system.

The equation of motion is obtained by considering Newton's second law $F = ma$ and the sum of forces acting on the mass $m$. For this case, the acting forces are those shown in Fig. 1.2, having:

$$-kx - cv + F(t) - mg + F_{est} = ma \tag{1.1}$$

In this formula, $F_{est}$ is the force produced by the mass weight $m$ ($g$, acceleration of gravity), and $a$ is the acceleration produced in the system, simplified in such a way as:

$$ma + kx + cv = F(t) \tag{1.2}$$

Considering the notation

$$v = \lim(\Delta x/\Delta t) \quad \Delta t \to 0; \quad dx/dt = \dot{x}$$
$$a = \lim(\Delta v/\Delta t) \quad \Delta t \to 0; \quad dv/dt = d^2/dt^2 = \ddot{x}$$

The equation of motion is

$$m\ddot{x} + c\dot{x} + kx = F(t) \tag{1.3}$$

This expression is the equation of motion of one degree of freedom with forced vibration and damping. The concept of "degree of freedom" refers to the minimum number of coordinates necessary to define the kinematics of a mechanical element. In this case, the system has one degree of freedom because it is only required to know the displacement $x$ to determine the position of the mass $m$.

The solution of the second-order differential equation has two elements, the homogenous solution and the particular solution. The homogenous solution represents the free vibration movement and the particular solution corresponds to the forced vibration of the mass.

The characterization of the oscillatory movement of the mass $m$ is obtained by defining the equation of movement according to the frequency ($f$) and the amplitude ($x$).

The oscillation frequency is the number of times per unit of time the movement repeats and its amplitude is the magnitude of the maximum displacement.

According to the previous, it can be inferred that $f = n/t$, where $n$ is the number of cycles completed in an interval $t$. If $n = 1$, it is said that $t = T$ is the oscillation period, meaning that $f = 1/T$ given in cycles/second (cycles/s) or Hertz, abbreviated with Hz. Due to the fact that a harmonic oscillation repeats regularly, it can be represented as shown in Fig. 1.3, where it can be seen that for a cycle to be complete, the period must be $T = 2\pi/\omega$. In this expression, $\omega$ is the angular frequency, that is, the angle that runs through the vector **A** per time unit. Taking into account the previously mentioned, the relation between $\omega$ and $f$ is given by $\omega = 2\pi f$.

Considering the amplitude and speed values with which the movement ($x_0$ y $v_0$) starts and which are generically known as initial conditions, the phase angle can be defined as $\phi = \tan^{-1}(v_0/x_0\omega)$.

Fig. 1.4 Illustrates the points of reference that define the peak amplitude, the peak to peak, the average value, and the root mean square (RMS) in a periodic signal and a random signal (without apparent order).

The peak-to-peak value indicates the end-to-end total displacement of the vibration. This value is required when, for example, with the maximum force in a machine part, it is presented with the maximum amplitude. The peak value is useful to indicate short-term impacts, without considering the history of the vibration. When it is required to know the average value of the vibration, the following expression is used:

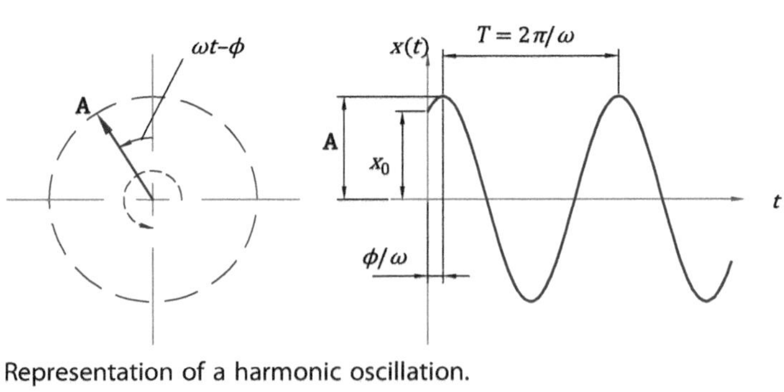


**Fig. 1.3** Representation of a harmonic oscillation.

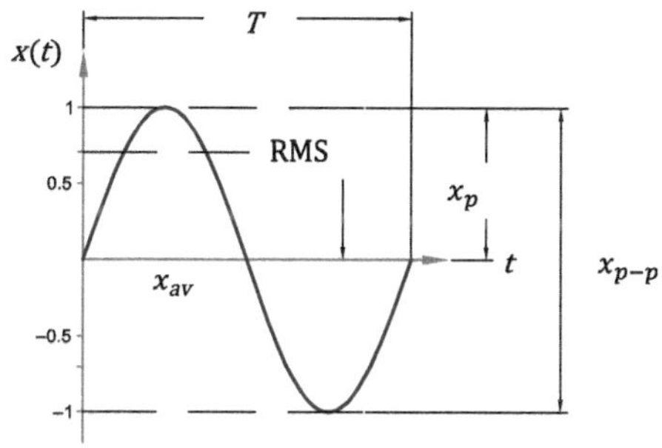


**Fig. 1.4** Definition of the peak values, peak to peak, average, and the root mean square in a vibration signal.

$$x_{prom} = 1/T \int_0^T x dt \tag{1.4}$$

The average of a sinusoidal function is always zero; therefore, its value has minimum contribution for the analysis of a vibration signal, and it is useless for a conditioning monitoring system. There is another measurement of the average amplitude of a sinusoidal signal, namely the root mean square (RMS), which is given by the following expression:

$$x_{RMS} = \left(1/T \int_0^T x^2 dt\right)^{1/2} \tag{1.5}$$

This formula is related to the evaluation of the vibration in the time and with the energy content of the vibratory wave. For a sine wave of unitary amplitude, the RMS value is 0.707 of the peak amplitude.

The simplest case is the free vibration of a particle without damping.

## Free vibration

Assuming there is an undamped motion with no external excitation, there would be an equation of the form $m\ddot{x} + kx = 0$ in such a way that:

$$\ddot{x} + \omega_n^2 = 0 \tag{1.6}$$

where

$$\omega_n = \sqrt{k/m} \tag{1.7}$$

This last expression is known as natural frequency. It is the frequency that the mass $m$ would oscillate after applying an excitation with minimum excitation energy and without damping.

The solution of the differential equation is of the form

$$x(t) = \alpha e^{\lambda t} \tag{1.8}$$

Using Euler's numbers, the solution of this equation can be transformed into

$$x = x_0 \cos(\omega_n t + \phi) \tag{1.9}$$

where $x_0$ and $\phi$ depend on the initial conditions.

At $t = 0, x = x_0, \dot{x} = \dot{x}_0$, therefore

$$\tan(\phi) = -\frac{\dot{x}_0}{x_0 \omega_n} \tag{1.10}$$

As can be seen, the free movement of the vibratory system is harmonic.

The natural frequency, which as a damping absence is equal to the resonance frequency, is a fundamental characteristic of the vibratory systems. Therefore, in a conditioning monitoring system, it is a fundamental parameter that can be determined from the vibration measurements because most of the vibratory effects being presented in the machines will be associated with the natural frequency. Even though the operating conditions of a machine are set away from the natural frequencies, it determines the amplitude response of a machine.

## Damped free vibration

In this case, the damping of the system is considered with the term $c\dot{x}$, where $c$ is the damping coefficient and $\dot{x}$ is the displacement speed. Considering again Fig. 1.3 and the sum of forces is:

$$m\ddot{x} + c\dot{x} + kx = 0 \tag{1.11}$$

that can be expressed as

$$\ddot{x} + (c/m)\dot{x} + (k/m)x = 0 \tag{1.12}$$

Using the definitions,

$$k/m = \omega_n^2 \tag{1.13}$$

$$\xi = c/2m\omega_n \tag{1.14}$$

with $\xi$ defined as the damping factor (the definition of this factor is derived from the solution of the second-order differential equation, and it is also known as the critical damping factor), the equation of motion remains as

$$\ddot{x} + 2\xi\omega_n^2\dot{x} + \omega_n^2 x = 0 \tag{1.15}$$

The solution of this equation depends on both the value of the damping factor $\xi$ and the initial conditions $x(0)$ and $\dot{x}(0)$. There are three possible solutions: overdamping ($\xi > 1$), underdamped ($\xi < 1$), and critical damping ($\xi = 1$).

In case the displacement and initial speed are $x(0) = 0$ and $\dot{x}(0) = v_0$, the solution of the equation for the overdamped case is:

$$x(t) = \frac{v_0}{\left(\xi^2 - 1\right)^{1/2}\omega_n} e^{-\xi\omega_n t}\operatorname{senh}\left[\left(\xi^2 - 1\right)^{1/2}\omega_n t\right] \tag{1.16}$$

Fig. 1.5 shows the variations as a function of the initial speed (Fig. 1.5A), the natural frequency (Fig. 1.5B), and the damping factor (Fig. 1.5C).

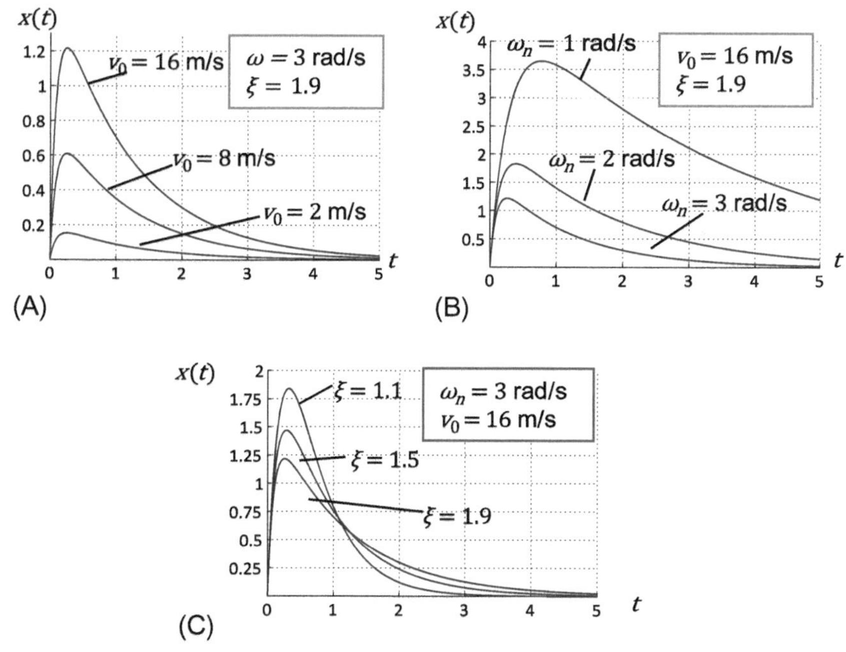


**Fig. 1.5** Variations of an overdamped vibration for different values of initial speed (A), natural frequency (B), and damping factor (C).

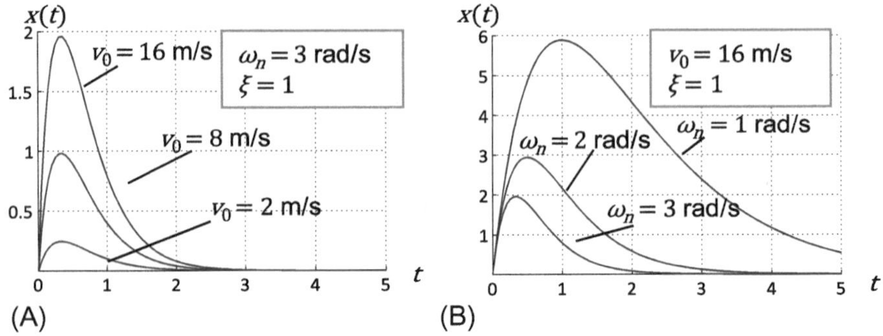


**Fig. 1.6** Variations of a critically damped vibration for typical values of initial conditions (A) and natural frequencies (B).

Fig. 1.6 shows the solution for the critical damping condition, where $x(t) = v_0 t e^{-\omega_n t}$ (this solution is a particular case for solving a second-order differential equation when the characteristic polynomial has two identical roots, also known as the critical damping condition). Fig. 1.6A represents the response when the system is excited at different initial conditions, and Fig. 1.6B shows the response of different systems with different natural frequencies.

For the underdamped case,

$$x(t) = (v_0/\omega_d)e^{-\xi\omega_n t}\mathrm{sen}\omega_d t, \tag{1.17}$$

with

$$\omega_d = \omega_n\sqrt{1-\xi^2} \tag{1.18}$$

This solution is found by solving a second-order differential equation with a characteristic polynomial with two complex roots. This is the most useful solution because it represents most of the dynamic problems found in mechanical systems. Fig. 1.7A shows the variations of this solution as a function in the initial speed, Fig. 1.7B corresponds to the solution as a function of the damping factor, and Fig. 1.7C shows the effect of different system parameters (different natural frequencies). The expression $\omega_d$ is known as the frequency of the damped free vibration, whose value will be lower when the damping factor of the system is higher.

The development of the previous equations is useful when identifying system parameters. The dynamic response of a mechanical system can be obtained by hitting the equipment with an instrumented hammer and measuring the vibrations with an accelerometer. The test results will produce a

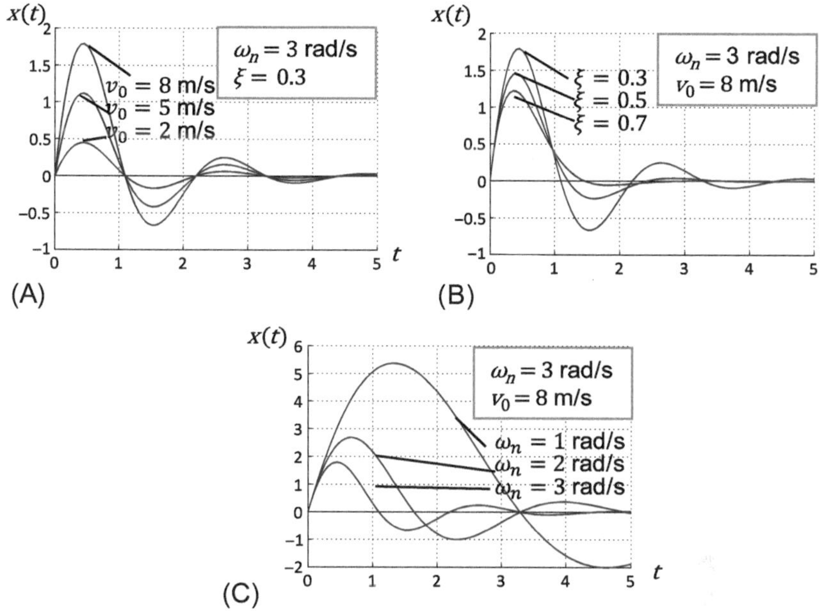


**Fig. 1.7** Variations of an undamped vibration for typical values of the initial speed (A), damping factor (B), and natural frequency (C).

similar graphic to those displayed in Fig. 1.7. From the analysis of the response, it is possible to evaluate the dynamic parameters, first determining the logarithmic decrement in order to find $\xi$. Fig. 1.8 presents the concept of the logarithmic decrement, which allows the damping of systems to be evaluated considering the decrease in amplitude between two consecutive cycles of a vibration. Considering the expressions $x(t)$ for a damped free vibration for either of the two consecutive cycles, if it takes place involving the times $t_1$ and $(t_1 + T)$, the following relation can be established:

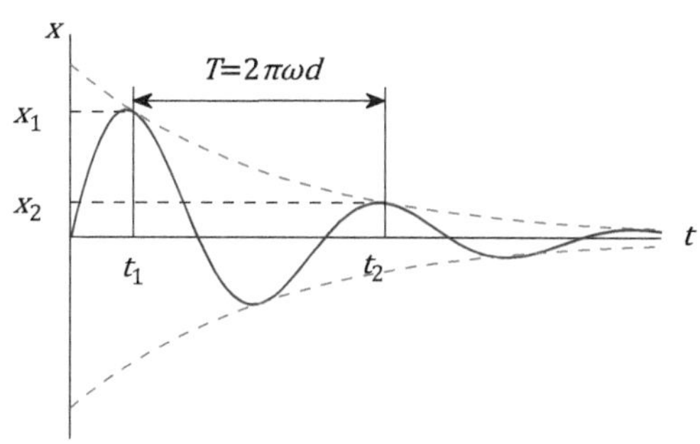


**Fig. 1.8** Logarithmic decrement.

$$\frac{x_1}{x_2} = \frac{e^{-\xi\omega_n t_1}}{e^{-\xi\omega_n (t_1 + T)}} = e^{\xi\omega_n T} \tag{1.19}$$

Where

$$\ln\frac{x_1}{x_2} = \xi\omega_n T = \delta \tag{1.20}$$

In this formula, $\delta$ represents the logarithmic decrement.

For the damped free vibration, $\omega_n$ is substituted by $\omega_n = \omega_d/\sqrt{1-\xi^2}$ and $T = 2\pi/\omega_d$, making it possible to establish the relation

$$\xi = \frac{\delta}{\sqrt{(2\pi)^2 + \delta^2}} \tag{1.21}$$

With the previous expressions, it is possible to obtain the damping of a system by measuring the amplitude of vibration of two consecutive cycles because knowing the logarithmic decrement can clear up the damping factor from the previous expression. In addition, by using the expression $\xi = c/2m\omega_n$ the damping coefficient $c$ can also be obtained. However, it's important to remark that the value of the damping coefficient of a system only depends on the damper while with the damping factor, it is determined by the parameters of the mechanical system.

## Forced vibration

Every machine is subjected to external excitations, which are related to the operating conditions and the mechanical configuration of each component. For conditioning monitoring systems, this characteristic boosts the prediction of failures at a component level. Thus, it is important to deeply understand the response of a mechanical system to a forced excitation. The simplest system is a one-degree-of-freedom mass-spring system (a simple linear pendulum of a single-degree-of-freedom oscillator has the same dynamic response). The most illustrative case is when the excitation force is represented as a simple harmonic excitation. The response of a one-degree-of-freedom system to a harmonic excitation is presented next. An essential part of the study of mechanical vibrations is the knowledge of the response of the system to an external excitation. In the case of the machinery, the main source of this external excitation comes from the power supply to the machine through the motors used for its operation. This means that, once the machine starts up, forced vibrations will occur. However, the total elimination of these vibrations is impossible because the very operation of the engines in industrial conditions is subject to the variation of their components, tolerances, mismatches, imbalances, variations in the power supply, and wear of parts; in other words, countless causes. The purpose, therefore, is to keep these vibrations at tolerable levels. With vibration monitoring for predictive maintenance, the intention is to identify the source of any change in the tolerable levels of forced vibration before they exceed the reference levels of normal operation.

Here, the response of a one-degree-of-freedom vibratory system to a harmonic excitement is studied, providing the basis for the spectral analysis application to the study of the forced vibration in machinery.

In the previous section, the solution for the homogenous equation, $m\ddot{x} + c\dot{x} + kx = 0$, was presented. This suggests the concept of the system response to an external excitation. However, it should be taken into account that the complete solution of the nonhomogeneous equation $m\ddot{x} + c\dot{x} + kx = F(t)$ is the sum of the solution of the homogeneous equation, called the transitory solution, plus the solution of the nonhomogeneous equation, called the permanent state solution, which will be present as long as the excitation is present $F(t)$.

From Fig. 1.4, the equation of motion is obtained by the sum of forces:

$$m\ddot{x} + c\dot{x} + kx = F(t) \tag{1.22}$$

According to this expression, $F(t)$, the force of external excitation, displaces the stiffness spring $k$ in a harmonic way. The excitation force is always proportional to the stiffness coefficient, thus:

$$F(t) = k \cdot f(t) = kA\cos(\omega t) \tag{1.23}$$

Here, $f(t)$ is the displacement function of the external force of the amplitude $A$ and the frequency $\omega$. Dividing between $m$ and considering the established relations $c/m$ and $k/m$ in the previous section, the equation of motion is:

$$\ddot{x} + 2\xi\omega_n\dot{x} + \omega_n^2 x = \omega_n^2 A\cos(\omega t) \tag{1.24}$$

For a harmonic motion, it is assumed that the solution of the previous equation is:

$$x(t) = X\cos(\omega t - \phi) \tag{1.25}$$

This approximation can be demonstrated from the particular solution of a second-order differential equation with a harmonic function. The demonstration is out of the scope of this chapter.

Here, $X$ represents the amplitude of the response to the force $F(t)$ and $\phi$ is the phase angle between the applied force and the resulting displacement of the system. The phase angle also represents the time delay between the excitation signal and the response signal. The importance of the property will be discussed latter because it is utilized for distinguishing the nature of two signals that have the same excitation frequency but different responses. $\dot{x}(t)$ and $\ddot{x}(t)$ are obtained deriving two times $x(t)$, and when substituted into the equation of motion, solving the algebra, the procedure generates two equations from which result the amplitude relation between the input $F(t)$ of amplitude $A$ and the output $x(t)$ of amplitude $X$ given by:

$$\frac{X}{A} = \frac{1}{\sqrt{\left[1-(\omega/\omega_n)^2\right]^2 + (2\xi\omega/\omega_n)^2}} = |H(\omega)| \tag{1.26}$$

The expression $|H(\omega)|$ is known as the magnification factor or the transfer function between the excitation force and the dynamic response.

In Fig. 1.9, the concept of phase angle is presented, indicating the period $T = 2\pi/\omega_n$ and phase $(t = \phi/\omega)$ between the signals $f(t)$ and $x(t)$. Therefore, for this case the phase angle between the input excitation force $F(t)$ and the resulting output displacement $x(t)$ is given by:

$$\phi = \tan^{-1}\frac{2\xi\omega/\omega_n}{1-(\omega/\omega_n)^2} \tag{1.27}$$

Fig. 1.10 shows the typical chart obtained from the relationship between the amplitude $X/A = |H(\omega)|$ and the frequency ratio $\omega/\omega_n$ for different values of the damping factor $\xi$. Fig. 1.11 shows the variations of the phase angle as a

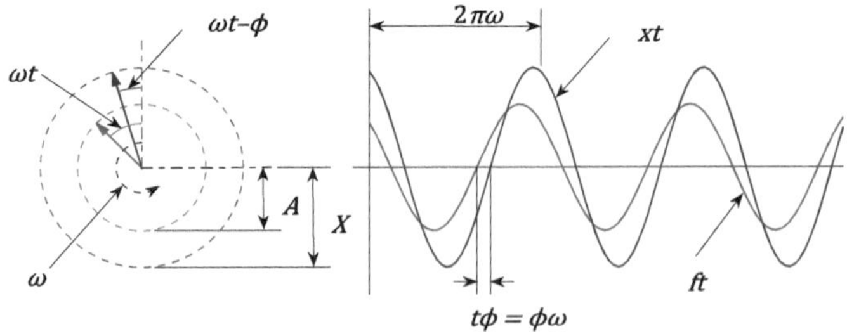


**Fig. 1.9** Phase angle between the excitation force and the output displacement.

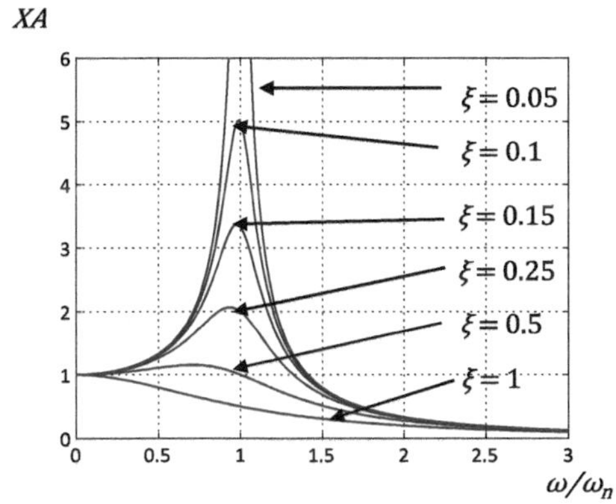


**Fig. 1.10** Magnification factor versus relation of frequencies.

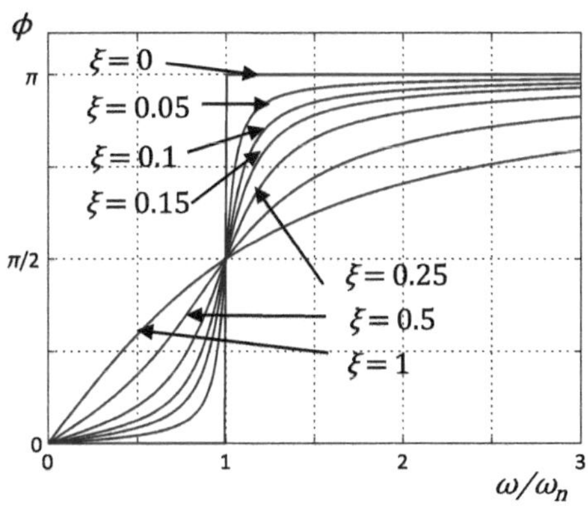


**Fig. 1.11** Phase angle versus relation of frequencies.

function of the frequency ratio and the damping factor. This analysis is very useful for understanding the transmissibility of the excitation forces within the machine components.

In the previous case, it was assumed that the base of the system was fixed. If it is considered that this base moves harmonically, the system can be represented by the scheme shown in Fig. 1.12, with the displacement and speed of the base given by $x_b(t) = A_1 \cos(\omega t)$ and $\dot{x}_b(t) = -A_1 \omega \mathrm{sen}(\omega t)$. With these, the equation of motion is expressed as follows:

$$\ddot{x} + 2\xi\omega_n\dot{x} + \omega_n^2 x = \omega_n^2 x_b + 2\xi\omega_n\dot{x}_b \tag{1.28}$$

The solution of this equation, as in the case seen in the previous section, is also of the form

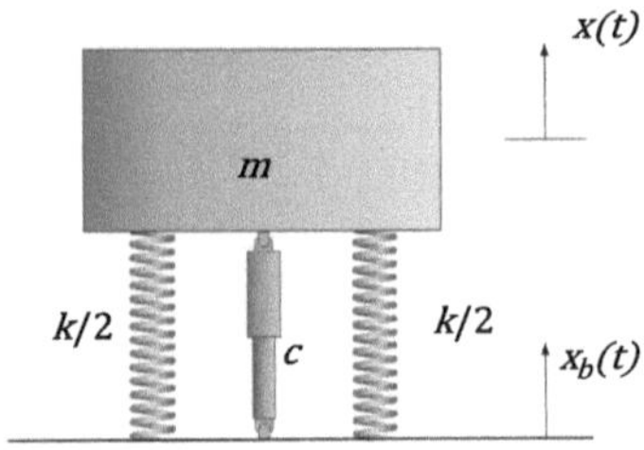


**Fig. 1.12** Vibratory system considering the motion of the base.

$$x(t) = X_1 \cos(\omega t - \phi_1) \tag{1.29}$$

If the base has harmonic excitation, the relation of the amplitudes is known as transmissibility and it is given by:

$$X_1/A_1 = |H(\omega)|\sqrt{1 + (2\xi\omega/\omega_n)^2} \tag{1.30}$$

The phase angle is:

$$\phi_1 = \tan^{-1}\frac{2\xi(\omega/\omega_n)^3}{1-(\omega/\omega_n)^2 + (2\xi\omega/\omega_n)^2} \tag{1.31}$$

This concept of transmissibility relates the force that the mass $m$ would transmit to the base where it is fixed, if in turn this mass would be experiencing a harmonic force $F_0(t) = F_0 \cos(\omega t)$.

Figs. 1.13 and 1.14 show the variations of the amplitude ratio and phase angle for different values of the damping coefficient. This analysis is known as the force transmitted by a machine to its base. This is, if the excitation force that the machine experiences is $F_0(t) = F_0 \cos(\omega t)$, the relation between the transmitted force ($F_{tr}$) and the applied force ($F_0$) to the base will be given by:

$$F_{tr}/F_0 = X_1/A_1 = |H(\omega)|\sqrt{1 + (2\xi\omega/\omega_n)^2} \tag{1.32}$$

Practically, it is impossible to eliminate the vibrations of a motor operating in industrial conditions due to manufacturing imperfections, component wear, and the interaction with other machines through its base. An example of the aforementioned is the case of an unbalanced motor that, when turning, has the shaft transmitting alternating forces to the basement. To analyze this case, it is important to consider the system shown in Fig. 1.15, which presents a mass motor $M$ with a support on its base, defined by a stiffness $k$ and a

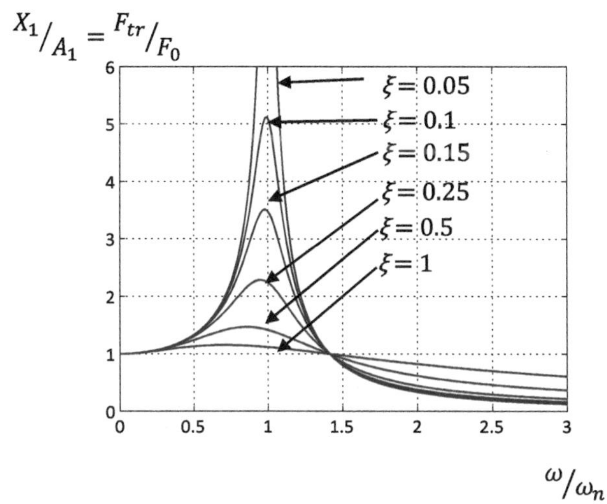


**Fig. 1.13** Variation of the amplitudes relation with the relation of frequencies, considering the harmonic motion of the base.

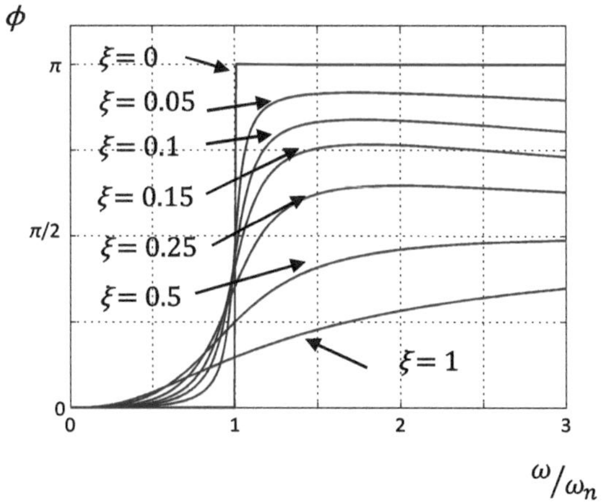


**Fig. 1.14** Variation of the phase angle with the relation of the amplitude, considering the harmonic motion of the base.

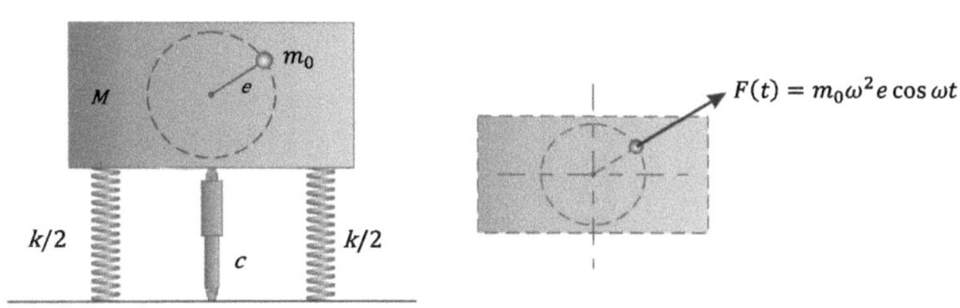


**Fig. 1.15** Representation of a motor transmitting alternating loads to its base in the vertical direction.

damping coefficient $c$. If it is considered that the imbalance of the system is caused by a mass $m_0$, which is a fraction of $M$ and spins at a distance $e$ around the rotating shaft of the motor, it can be established that, in the vertical shaft, the force that produces the imbalance is $F(t) = m_0\omega^2 e \cdot \cos(\omega t)$. In this case, the lateral motion that would produce the horizontal component of the imbalance force is ignored. If required, it would be made in the horizontal direction, a similar approach to the one conducted for the vertical motion of the motor.

It is $F_0 = m_0\omega^2 e$, by which the relationship between the transmitted force and the present force due to imbalance is:

$$F_{tr}/\left(m_0\omega^2 e\right) = |H(\omega)|\sqrt{1 + (2\xi\omega/\omega_n)^2} \tag{1.33}$$

It must be noted that $F_0$ is not constant but instead depends on the frequency $\omega$, which is the angular speed of the motor shaft, dividing both sides of Eq. (1.33) by $m_0\omega_n^2 e$ the relation of the normalized transmitted force $F_{tr}$ is obtained, normalized in respect to the transmitted force to the natural frequency, that is, the force that would be transmitted with the motor shaft spinning to its natural frequency. In this way, the previous relation is:

$$F_{tr}/\left(m_0\omega_n^2 e\right) = (\omega/\omega_n)^2|H(\omega)|\sqrt{1 + (2\xi\omega/\omega_n)^2} \tag{1.34}$$

Fig. 1.16 presents the variation of the $F_{trMAX}$ with respect to the relation of frequencies $\omega/\omega_n$. In this figure, it can be seen that in the case of the

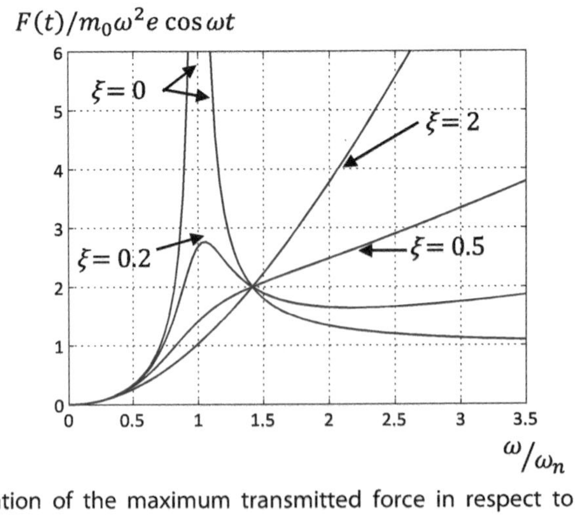


**Fig. 1.16** Variation of the maximum transmitted force in respect to the relation of frequencies.

resonance, that is, if $\omega/\omega_n$ and if the damping factor is $\xi=0$, the transmitted force tends to infinite. It is also interesting to note that the lower the damping factor, the greater the transmitted force.

However, if $\xi$ is bigger, the transmitted force will increase with the frequency, indicating a clear need to analyze in detail and with solid foundations each vibration problem of the machinery. In many cases, increasing the damping factor of a machine not only does not solve the vibration problem but actually aggravates it.

## Transmissibility

The definition of the previous section is very useful when designing the supports of a machine, the shock absorbers or the foundation. To design a shock absorber, for example, it is necessary to specify its stiffness and damping coefficients. To find these values, it is important to determine certain design parameters from the transmissibility function, which is defined as:

$$T_f = \sqrt{\frac{1+4\zeta^2\left(\frac{\omega}{\omega_n}\right)^2}{\left(1-\left(\frac{\omega}{\omega_n}\right)^2\right)^2+4\zeta^2\left(\frac{\omega}{\omega_n}\right)^2}} \tag{1.35}$$

In this case, the designer must establish a certain transmissibility value and a specific frequency. Fig. 1.17 shows the transmissibility as a function of $\omega/\omega_n$. It is impossible to have zero transmissibility but an acceptable value is below one. From Fig. 1.17 it is clear to find that $\omega/\omega_n > \sqrt{2}$.

Once the design conditions are determined and the material properties selected, the geometry of the absorber must be defined.

To better understand the effect that the excitation force has on the speed of the equipment, it is necessary to define the concept of critical speed.

## Critical speed

One concept of the vibration theory that has major application in the analysis of the rotating machinery is the critical speed of a rotor. This is because, at these speeds, the amplitudes of maximum vibrations are presented; these occur at the natural frequencies of the rotor. When the angular frequency of rotation of the rotor ($\omega$) is equal to the natural frequency of the aforementioned ($\omega_n$), it is said that it goes through a critical speed of the

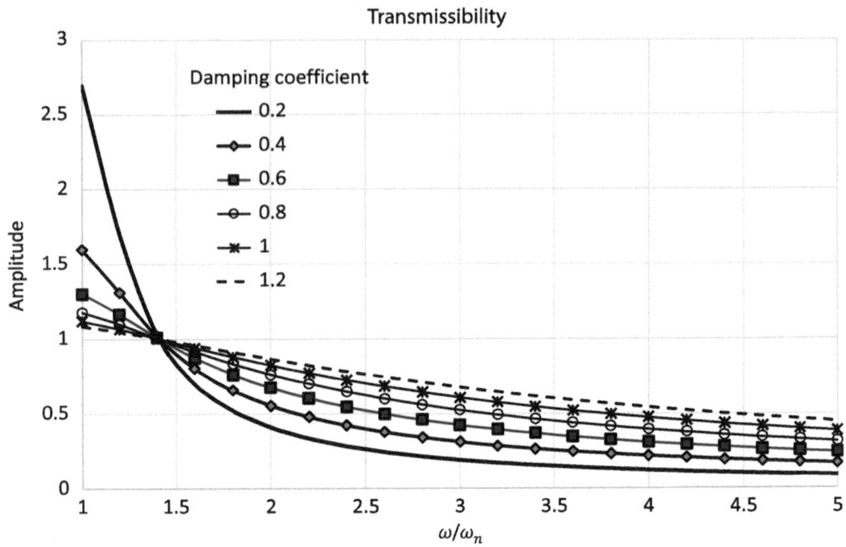


**Fig. 1.17** Transmissibility as a function of the excitation frequency.

rotor, meaning it is in resonance. The operating speed of a machine must be separated from the critical speeds because if $\omega$ is similar to $\omega_n$, then there will be vibrations of great amplitude leading to high intensity efforts, possible friction between the rotating parts, and the transmission of harmful forces to the foundation of the machine.

The problem of the critical speed of a rotor has diverse causes, especially the ones associated with a nonhomogeneous distribution of the system's mass, hysteresis, gyroscopic effects, and bearings.

In this section, the basis for the analysis of the problem is described, considering only one nonhomogeneous mass distribution and assuming that the one of the rotor is negligible compared to that of a disc that concentrates the mass of the system mass located in the center of the clearance between the bearings, as shown in Fig. 1.18. In this system, the center of the system $G$ does not match with the geometric center $A$ of the disc, which causes the imbalance. The distance $AG$ is known as the eccentricity ($e$) and the center of rotation in the position of equilibrium is $O$.

Assume that the system rotates at an angular speed $\omega$, that all damping forces are proportional to the speed of the geometric center of the disc, and that the bearings are perfectly rigid. It can then be established that the damping force $F_a$ and the elastic $F_e$ are given as this:

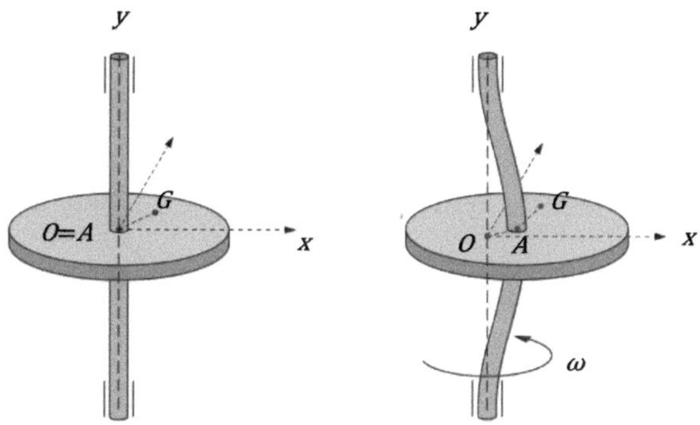


**Fig. 1.18** Rotating system considering that the mass of the system is concentrated in a disc at the center of the clearance of the bearing.

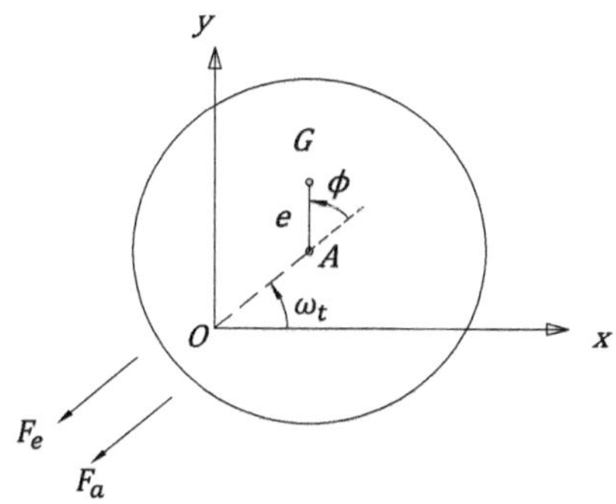


**Fig. 1.19** Plane *xy* of the disc shown in Fig. 1.18.

$$[F_a] = \begin{bmatrix} -c\dot{x} \\ -c\dot{y} \end{bmatrix}; \quad [F_e] = \begin{bmatrix} -kx \\ -ky \end{bmatrix} \tag{1.36}$$

In the $xy$ plane, the disc is represented as shown in Fig. 1.19 where the acting forces are indicated, meaning the rotation in respect to the geometric center and the phase angle. Taking into account the established in the section of forced vibration, it can be established that the equations of motion are:

$$m\ddot{x}_A + c\dot{x}_A + kx_A = me\omega^2 \cos(\omega t) \tag{1.37}$$

$$m\ddot{y}_A + c\dot{y}_A + ky_A = me\omega^2 \text{sen}(\omega t) \tag{1.38}$$

Having:

$$x_A = \frac{me\omega^2 \cos(\omega t - \phi)}{\sqrt{(k - m\omega^2)^2 + (c\omega)^2}}; \quad y_A = \frac{me\omega^2 \text{sen}(\omega t - \phi)}{\sqrt{(k - m\omega^2)^2 + (c\omega)^2}} \tag{1.39}$$

From the previous relations, the radial displacement, as a function of the natural frequency, can be found as:

$$\overline{OA} = \frac{e(\omega/\omega_n)^2}{\sqrt{\left[1-(\omega/\omega_n)^2\right]^2 + (2\xi\omega/\omega_n)^2}} \tag{1.40}$$

The phase angle is:

$$\phi = \tan^{-1}\frac{2\xi(\omega/\omega_n)}{1-(\omega/\omega_n)^2} \tag{1.41}$$

According to the previous expression, three cases for the relative position of the points $O$, $A$, and $G$ can be identified as follows: If si $\omega < \omega_n$, $\phi < 90°$; when $\omega = \omega_n$, $\phi = 90°$; with $\omega > \omega_n$, $\phi > 90°$.

It is important to remember that for an angular speed of determined operation, the phase angle is defined in a unique way. In addition to the fact that when the angular speed of operation is higher than the critical speed, the center of mass $G$ tends to rotate at the same angular velocity as the center of rotation $O$.

In some machines, the operating speed is higher than the natural frequency. This means that during starting and stopping operations of the machine, it goes through a resonance. If the angular speed of the machine presents values close to the critical speed, vibrations of great amplitude will be produced. However, if it goes through this critical value with enough speed, the amplitude of vibration will not increase to dangerous levels.

The critical angular speed is also known as whirl. The stable condition occurs when

$$\dot{\phi} = \omega \tag{1.42}$$

Integrating this equation:

$$\phi = \omega t - \varphi \tag{1.43}$$

## System with two or more degrees of freedom

Recalling the definition of degrees of freedom, it is necessary to entirely know the motion of the vibratory system shown in Fig. 1.20. It is required to know the variations of $x_1(t)$ and $x_2(t)$. The system has

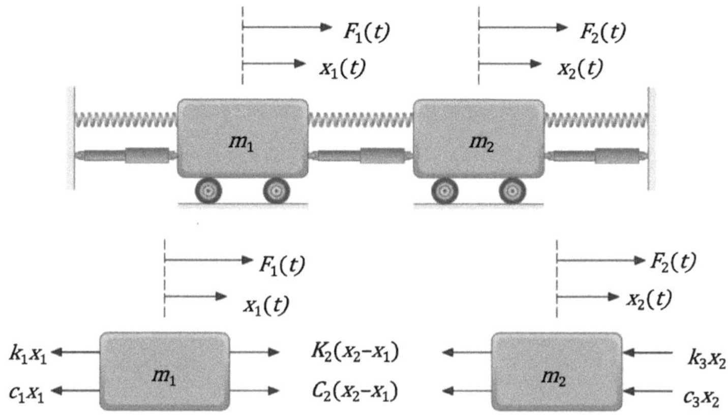


**Fig. 1.20** Two-degree-of-freedom vibration system.

two degrees of freedom and two conditions of resonance, that is, two natural frequencies. In the case of rotating systems, generally it can be said that they have two critical speeds. In each resonance condition, the system will oscillate in a particular way, called a vibration mode. A two-degree-of-freedom system also has two vibration modes that occur at their natural frequencies. These modes are the configuration that the system takes when vibrating to each of its resonance frequencies. The modes are also found from the solution of the eigenvectors of the fundamental equation of motion.

The study of vibration systems with two degrees of freedom allows the generalization of the analysis methods. This analysis can be applied to systems with multiple degrees of freedom. The number of natural frequencies and vibration modes is equal to the number of degrees of freedom. Generally, the degree of freedom of a system will be equal to the number of concentrated masses of the same one.

The equations of motion of a system with $n$ degrees of freedom will be a system of simultaneous ordinary differential equations. That is, the motion of a mass will depend on the motion of the other. However, by properly selecting the reference system, the main or natural coordinates can be established. By using these coordinates, the differential equations of the system become independent and with a similar structure to those systems with one degree of freedom. Likewise, the principle of superposition can also be applied, as it allows each mode of a vibration system to be analyzed in an independent way to then add its effects to know the total response of the system.

The general methodology to establish the equations of motion of a vibration system with two degrees of freedom is presented below. For the system shown in the previous figure, the sum of forces on the masses is:

$$F_1(t) - c_1\dot{x}_1 - k_1x_1 + c_2(\dot{x}_2 - \dot{x}_1) + k_2(x_2 - x_1) = m_1\ddot{x}_1 \quad (1.44)$$

$$F_2(t) - c_2(\dot{x}_2 - \dot{x}_1) - k_2(x_2 - x_1) - c_3\dot{x}_2 - k_3x_2 = m_2\ddot{x}_2 \quad (1.45)$$

Rearranging these equations, the result is:

$$m_1\ddot{x}_1 + (c_1 + c_2)\dot{x}_1 - c_2\dot{x}_2 + (k_1 + k_2)x_1 - k_2x_2 = F_1 \quad (1.46)$$

$$m_2\ddot{x}_2 - c_2\dot{x}_1 + (c_2 + c_3)\dot{x}_2 - k_2x_1 + (k_2 + k_3)x_2 = F_2 \quad (1.47)$$

In matrix code:

$$\begin{bmatrix} m_1 & 0 \\ 0 & m_2 \end{bmatrix}\begin{Bmatrix} \ddot{x}_1 \\ \ddot{x}_2 \end{Bmatrix} + \begin{bmatrix} c_1 + c_2 & -c_2 \\ -c_2 & c_2 + c_3 \end{bmatrix}\begin{Bmatrix} \dot{x}_1 \\ \dot{x}_2 \end{Bmatrix} + \begin{bmatrix} k_1 + k_2 & -k_2 \\ -k_2 & k_2 + k_3 \end{bmatrix}\begin{Bmatrix} x_1 \\ x_2 \end{Bmatrix} = \begin{Bmatrix} F_1 \\ F_2 \end{Bmatrix} \quad (1.48)$$

Defining $\begin{bmatrix} m_1 & 0 \\ 0 & m_2 \end{bmatrix} = [m]$ as the matrix of the mass, $\begin{bmatrix} c_1 + c_2 & -c_2 \\ -c_2 & c_2 + c_3 \end{bmatrix} = [c]$ as the damping matrix, $\begin{bmatrix} k_1 + k_2 & -k_2 \\ -k_2 & k_2 + k_3 \end{bmatrix} = [k]$ as the stiffness matrix, and $\begin{Bmatrix} x_1 \\ x_2 \end{Bmatrix} = \{x\}$ and $\begin{Bmatrix} F_1 \\ F_2 \end{Bmatrix} = \{F\}$, as the displacement and force vectors, respectively, the equation of motion can be described as

$$[m]\{\ddot{x}\} + [c]\{\dot{x}\} + [k]\{x\} = \{F\} \quad (1.49)$$

As of this matrix equation, the free and forced vibration cases with harmonic excitation can be analyzed, with or without damping.

The solution of this ordinary differential equation system for the diverse mentioned cases involves the use of mathematical methods that are outside this work's scope but can be consulted in the references at the end of this book.

Following the objective to present the basic concepts of the theory of vibrations to its correct application to predictive maintenance, as described below, it is the case of the undampened free vibration of a system with three degrees of freedom.

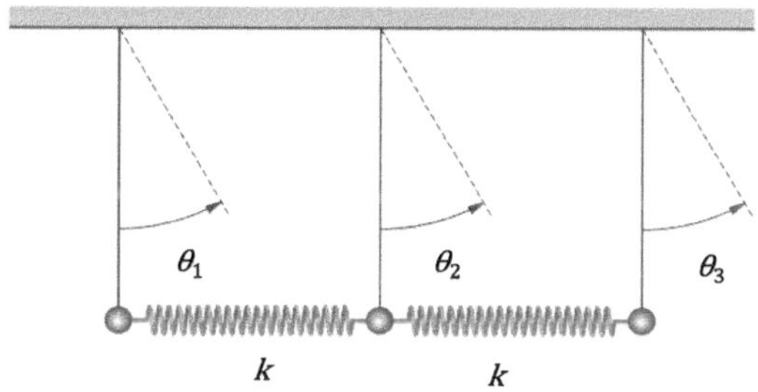


**Fig. 1.21** System with three degree of freedom.

Consider the system of a triple pendulum shown in Fig. 1.21, where the coordinates of the motion are indicated as $\theta_1$, $\theta_2$, and $\theta_3$, and suppose the three masses of equal $m$ value, with the springs defined by their constant of stiffness $k$.

For this case, the equation of motion is given by the expression:

$$\{\ddot{\theta}\} + [D]\{\theta\} = \{0\} \tag{1.50}$$

where $[D] = \begin{bmatrix} \frac{g}{l} + \frac{k}{m} & -\frac{k}{m} & 0 \\ -\frac{k}{m} & \frac{g}{l} + 2\frac{k}{m} & -\frac{k}{m} \\ 0 & -\frac{k}{m} & \frac{g}{l} + \frac{k}{m} \end{bmatrix}$, $\{\theta\} = \begin{Bmatrix} \theta_1 \\ \theta_2 \\ \theta_3 \end{Bmatrix}$ and $\{\ddot{\theta}\} = \begin{Bmatrix} \ddot{\theta}_1 \\ \ddot{\theta}_2 \\ \ddot{\theta}_{3.} \end{Bmatrix}$

Supposing harmonic solutions of the form $\theta_i = \Theta_i \cos(\omega t - \phi_i)$ with $i$ varying from 1 to 3, when substituted in the equation of motion, the following problem of characteristic values can be presented:

$-\omega^2\{\Theta\} + [D]\{\Theta\} = \{0\}$ which is reduced to $[D - \omega^2[I]]\{\Theta\} = \{0\}$

Here $[I]$ is a unitary triangular matrix.

The problem of characteristic values consists of finding the values of $\omega^2$, which allow meeting the previous equation. It is necessary that the determinant of the matrix be equal to zero. So, it is said that the characteristic equation of the vibration system has been obtained with the following expression:

$$\det\left[D - \omega^2[I]\right] = \{0\} \tag{1.51}$$

When solving the characteristic equation, the three natural frequencies are:

$$\omega_1 = \left(\frac{g}{l}\right)^{1/2}, \quad \omega_1 = \left(\frac{g}{l} + \frac{k}{m}\right)^{1/2}, \quad \omega_1 = \left(\frac{g}{l} + 3\frac{k}{m}\right)^{1/2} \tag{1.52}$$

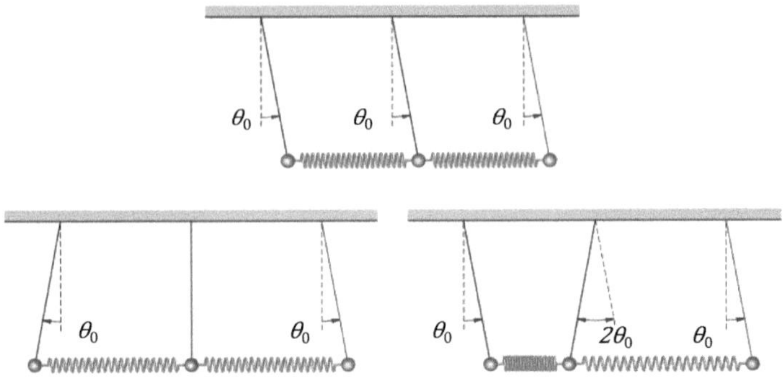


**Fig. 1.22** Modes of vibration of a triple pendulum shown in Fig. 1.21.

By substituting consecutively the values of each of the natural frequencies in the equation that presents the problem of the characteristic values, the following can be observed:

$$\left[D-\omega^2[I]\right]\{\Theta\}=\{0\} \tag{1.53}$$

The characteristic vectors (eigenvectors) are obtained $\{\Theta\}_i$,which define the modes of vibration.

In respect to the case presented here, for $\omega_1$, the first mode of vibration is defined by the vector $\{\Theta\}_1=\{1 \;\; 1 \;\; 1\}$; for $\omega_2$, the vector that characterizes the second mode is $\{\Theta\}_2=\{1 \;\; 0 \;\; -1\}$; and for $\omega_3$, the third mode of vibration is given by $\{\Theta\}_3=\{1 \;\; -2 \;\; 1\}$.

Fig. 1.22 presents the configuration of this vibration system in each of its modes of vibration.

It can be seen that in the first mode, the three bars oscillate synchronously to the frequency $\omega_1$ with the same amplitude. In the second mode, the central bar has no motion while those on the other end do it symmetrically at the frequency $\omega_2$. In the third mode of vibration, the central bar oscillates at the frequency $\omega_3$ with an amplitude twice the size compared to those of the side bars, which have a synchronous motion to the same frequency.

## Continuous systems

For analysis purposes, the vibration systems are modeled as concentrated masses in which the elasticity is also supposed to be acting at one point. This provides good results in a great number of cases. An alternative method of modeling consists of distributing both the mass and the elasticity of the

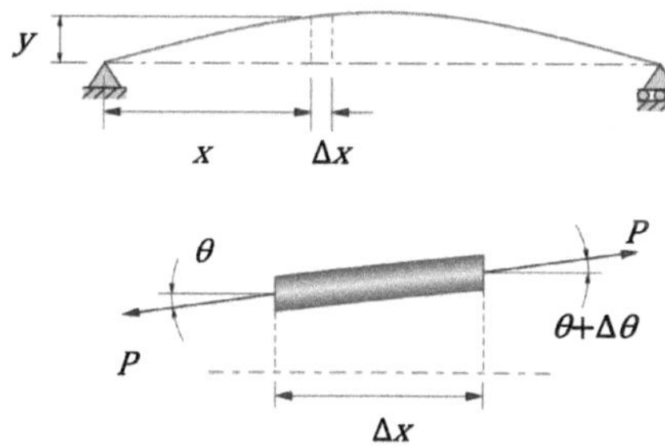


**Fig. 1.23** String vibrating transversally represented as a continuous system.

system, which happens to be very suitable in the analysis of strings, beams, rotors, and cables. The fundamental difference between the discrete and continuous models is that the first one results in ordinary differential equations while the second one is modeled with partial differential equations. This defines differences in the solution methods.

In this section, just the problem of undamped free vibration of one string is presented, focusing on two objectives: The analysis of this problem includes the study of the relationship between the tension and the natural frequency. This phenomenon occurs in applications such as cables or bands.

Fig. 1.23 presents a string subject to tension to which its uniformly distributed mass is considered. It is important to note that the transverse vibration $y$ depends on the values of the coordinate $x$ and the time $t$. Because of this, the derivatives established in the equation of motion will be impartial.

By making a sum of forces in the direction $y$, the result is:

$$-P\mathrm{sen}\theta + P\mathrm{sen}(\theta + \Delta\theta) = \rho\left(\partial^2 y/\partial t^2\right) \tag{1.54}$$

Here, $P$ is the tension of the string, $\rho$ is its mass per unit of length, and $y = y(x, t)$ is the transversal displacement of the string. This displacement $y$ varies according to the position of the analyzed point (defined by $x$) and of time.

By using the trigonometric relation $\mathrm{sen}(\theta + \Delta\theta) = \mathrm{sen}\theta\cos(\Delta\theta) + \cos\theta\,\mathrm{sen}(\Delta\theta)$ and the notation $(\partial^2 y/\partial x^2) = y''$ and $\left(\partial^2 y/\partial t^2\right) = \ddot{y}$, the equation of motion for this case is

$$y''c^2 = \ddot{y} \tag{1.55}$$

The expression $= \sqrt{P/\rho}$ is defined as the speed of the wave.

To obtain the solution of this equation, it is necessary to establish what are known as the initial conditions and the boundary conditions. The boundary conditions define the position and characteristics of the system

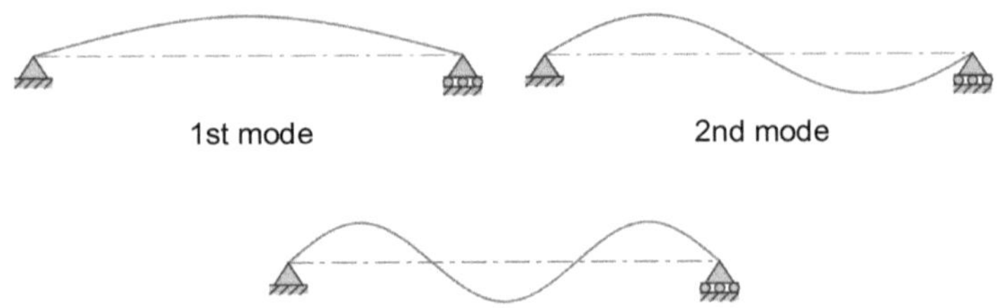


**Fig. 1.24** Modes of vibration of a string subject to its opposite ends.

in its edges. For this case, the boundary conditions are $y=0$ for $x=0$ and $x=l$. That is, the transversal displacement of the string is zero on the other ends.

With this, the solution of the equation is:

$$y(x,t) = (A_1 \mathrm{sen}\omega_n t + A_2 \cos \omega_n t)\mathrm{sen}(\omega_n x/c) \tag{1.56}$$

The constants $A_1$ y $A_2$ depend on the initial conditions, that is, those that establish the speed and acceleration of the system at the beginning of the time interval under study. From the previous expression, it can be seen that for the boundary conditions established, $y(x, t)$ will be zero for $x=0$ or $x=l$, if $\mathrm{sen}(\omega_n x/c)=0$. The preceding is met when the following occurs:

$$\omega_n = \frac{\pi n c}{l} = \pi n \left(\frac{P}{\rho l^2}\right)^{1/2} \tag{1.57}$$

The previous expression presents the natural frequency of the system under study and allows the tension of a string to be obtained through a simple free vibration test. Fig. 1.24 presents the configuration that a string must have for its first three modes of vibration.

CHAPTER TWO

# Spectral analysis

## Introduction

In general, the vibrations of a machine produce complex signals that do not have a unique sinusoidal shape, but are composed of various signals correlated with each other with different frequency amplitudes and phases. Furthermore, there are vibrations whose waveform has no apparent order. It is this correlation between signals that makes necessary its analysis through mathematical manipulations that allow handling them systematically. The main tool for the application of vibration analysis to predictive maintenance is the so-called spectral analysis based on the separation of the harmonic components of the vibratory signal, in such a way that this separation allows identifying the causes and effects of vibrations present in the operation of a machine. Spectral analysis is based on the Fourier series and the so-called fast Fourier transform (FFT). In an industrial environment, this analysis is carried out with the application of electronic instruments and specialized computer equipment. In order to use the equipment and instruments with all their potential, it is necessary to know the basis of the spectral analysis. This chapter presents the basic theory of Fourier's analysis and its application to the vibration analysis.

## Types of signals

In vibration analysis, it is possible to identify three types of signals: harmonic, periodic, and nonperiodic. The latter are also known as transitory. A common characteristic of these signals is that their frequency and amplitude are defined for any $t$ time. These types of signals are known as deterministic functions. In Fig. 2.1, a harmonic, a periodic, and a transitory signal are shown. The latter does not vary regularly with time; however, every time it occurs, its waveform will be defined. Such is the case of the start of an engine, which may not occur cyclically but when it happens, its waveform is determined.

*Mechanical Vibrations and Condition Monitoring*
https://doi.org/10.1016/B978-0-12-819796-7.00002-0

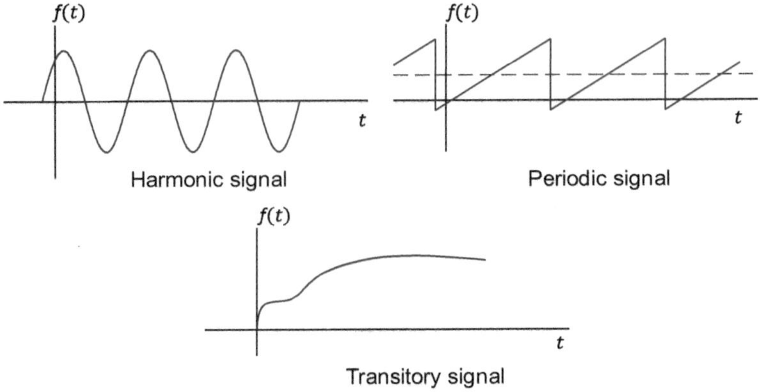


**Fig. 2.1** Deterministic signals: Harmonic, periodic, and nonperiodic or transitory.

Some vibration phenomena cannot be characterized because the frequency and amplitude vary with time. This means that if the frequency and amplitude of the signal under study were measured, there would not be enough information to define this signal. For example, measuring the vibration present in a trailer truck gearbox circulating on a highway does not clearly define how such an event is going to happen if this vehicle goes back on the same road again. These types of nondeterministic signals are known as *random vibration*.

Despite their great variability, the vibrations to be studied in the predictive maintenance field present a certain statistical regularity. Therefore, by considering a certain number of measurements and by applying a statistical criterion, this can characterize the vibratory system under analysis. Fig. 2.2 presents a random signal and the measurement process to obtain some values that would show a certain average regularity.

## Time and frequency domains

Depending on the type of signal present in the vibration of certain machinery, the analysis might vary in complexity. If there were a pure harmonic signal, the determination of the frequency and amplitude would be relatively simple, and thus the cause and effect produced by this vibration could be identified. This identification can be carried out by applying the concepts covered in Chapter One. That is, by applying the basic theory of systems with excitation and harmonic responses. However, this does

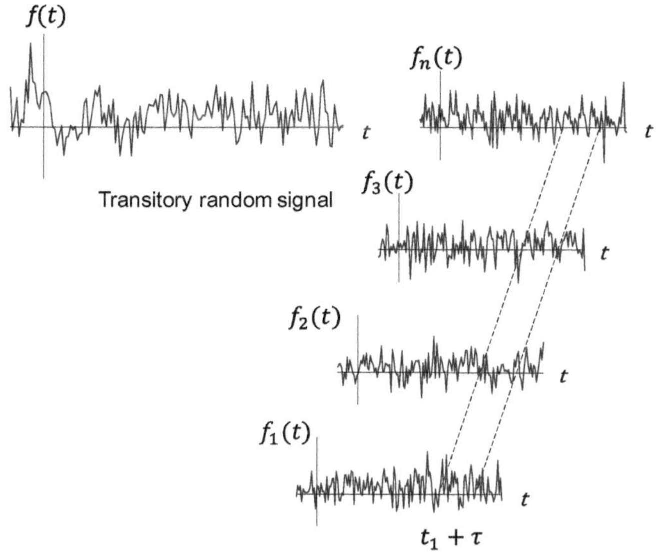


**Fig. 2.2** Random signal and averaging signal processing.

not usually occur in practice but instead random, periodic, or transitory signals are presented.

The great advantage of using spectral analysis or frequency analysis in the predictive maintenance of machinery is that it allows the decomposition of any signal into its harmonic components. This decomposition is achieved by applying the Fourier series concept to the signal under study. If there were a complex signal $f(t)$ in the time domain, that is, represented in the amplitude/time plane, such as the one presented in Fig. 2.2, a series of harmonics of different amplitudes and frequencies can be obtained in such a way that the sum of all these harmonic signals gives the initial signal as a result. If the different components of the complex signal are put into a graph, taking into account the frequency and amplitude of each of them in the frequency domain, that is, in an amplitude-frequency plane, it would have the frequency spectrum of the analyzed signal.

Fig. 2.3 shows examples previously described. In the first case, Fig. 2.3 presents a tuning fork that, when struck, produces a pure harmonic vibration. In respect to the time domain, this signal is graphically represented as a sinusoidal one of amplitude $A$ and frequency $\omega$. Considering the frequency domain, this signal is graphically represented with a single peak of frequency $\omega$ and amplitude $A$. The case study of a gear transmission represents a much more complex signal. In the time domain, this signal does not seem to have

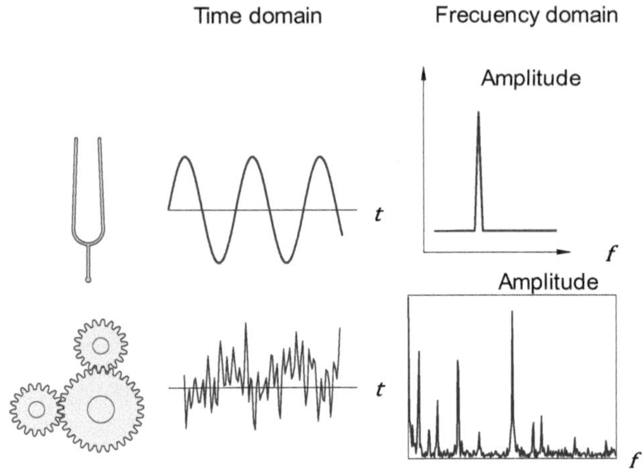


**Fig. 2.3** Examples of signals in the time and frequency domains.

any order. However, within the frequency domain, it is possible to identify defined peaks of frequency and amplitude related to the different causes and effects of the gear's vibrations.

Fig. 2.4 presents an example that illustrates in detail the passing of the time domain to the frequency domain. Within the time domain, the piston motion is represented by two harmonic waves of different amplitude and frequency, ordered in such a way that, added point to point, reproduce exactly the total motion waveform. In the frequency domain, the total motion of this piston would be represented as a signal with two peaks of frequencies, $\omega_1$ and $\omega_2$, with amplitudes $B_1$ and $B_2$, respectively, corresponding to the frequency and amplitude of the harmonic components from the motion produced by the piston.

## Fourier series

The frequency analysis of the vibration signals is based on the so-called Fourier series, which allows any vibratory signal to be decomposed into its harmonic components.

The mathematical method was proposed by Jean-Baptiste-Joseph Fourier in 1822, in relation to the analysis of heat transfer problems. Currently, this method is an indispensable tool in any problem of modern physics, vibration analysis, communications theory, or linear system. The Fourier series can be expressed in the form of trigonometric functions or exponential ones. In their trigonometric function form, the Fourier series is represented as follows:

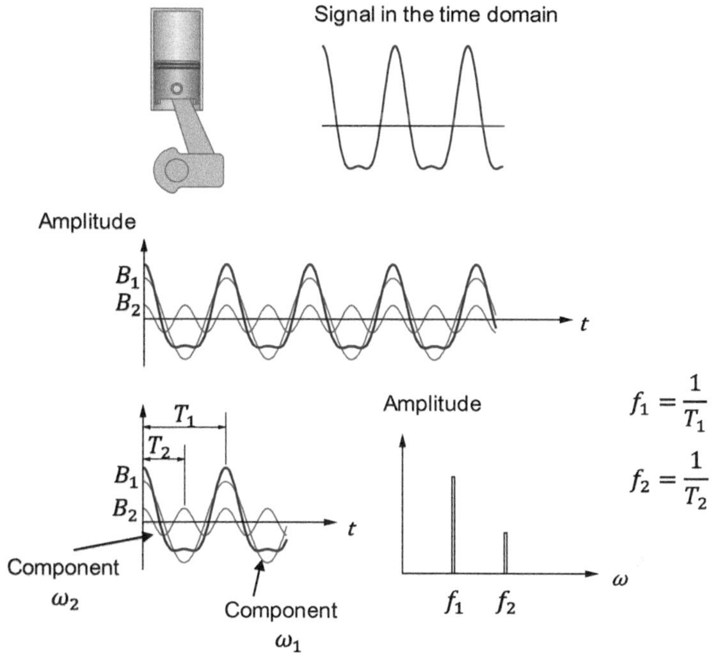


**Fig. 2.4** Representation of a signal transformation from the time domain to the frequency domain.

$$f(t) = \frac{1}{2}a_0 + \sum_{n=1}^{\infty} (a_n \cos n\omega_0 t + b_n \operatorname{sen} n\omega_0 t) \tag{2.1}$$

This can also be expressed as:

$$f(t) = C_0 + \sum_{n=1}^{\infty} C_n \cos (n\omega_0 t - \theta_n) \tag{2.2}$$

With the coefficients $a_0$, $a_n$, $b_n$, $C_0$, and $C_n$ defined as follows

$$a_0 = \frac{2}{T}\int_{-T/2}^{T/2} f(t)dt, \quad a_n = \frac{2}{T}\int_{-T/2}^{T/2} f(t)\cos(n\omega_0 t)dt, \\ b_n = \frac{2}{T}\int_{-T/2}^{T/2} f(t)\operatorname{sen}(n\omega_0 t)dt \tag{2.3}$$

$$C_0 = \frac{1}{2}a_0 \; y \; C_n = \sqrt{a_n^2 + b_n^2} \tag{2.4}$$

The phase angle is $\theta_n = \tan^{-1}(b_n/a_n)$.

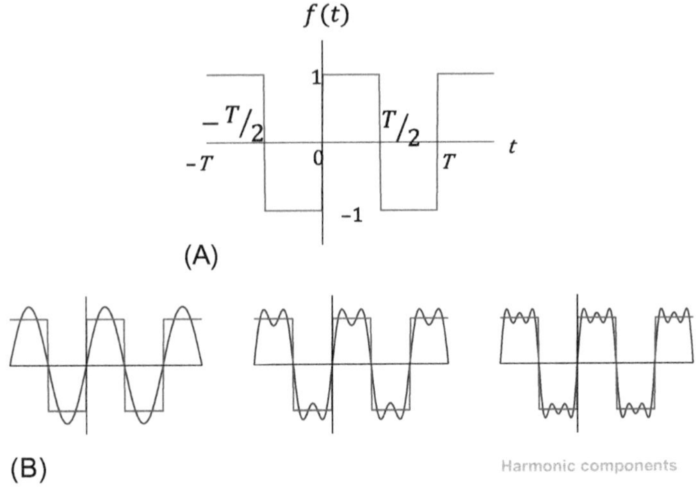


**Fig. 2.5** Representation of a periodic function by means of its harmonic components. (A) Original function, (B) approximations with 1, 2, and 3 terms.

In the previous expressions, $f(t)$ represents the function that, in the time domain, contains the total vibratory signal under study and analysis. $\omega_0$ designates the fundamental frequency and $n=1, 2\ldots$, indicates the $\omega_n$ frequencies that, along with the fundamental one, form the original signal.

As an example of the application of the previous expressions, in Fig. 2.5A a square wave is shown that, in the time domain, is defined by $f(t)=1$ if $t$ is greater than zero, but less than $T/2$, and $f(t)=-1$ if $t$ is bigger than $T/2$, but less than $T$. These conditions are normally expressed as follows:

$$f(t)=\begin{cases} 1, & 0<t<\dfrac{T}{2} \\ -1, & \dfrac{T}{2}<t<T \end{cases} \tag{2.5}$$

Applying the previous expressions, defined by this function, the result is

$$f(t)=\frac{4}{\pi}\left(\operatorname{sen}\omega_0 t+\frac{1}{3}\operatorname{sen}3\omega_0 t+\frac{1}{5}\operatorname{sen}5\omega_0 t+\cdots\right) \tag{2.6}$$

In the alternate trigonometric form, the above-mentioned function would be

$$f(t)=\frac{4}{\pi}\left[\cos\left(\omega_0 t-\frac{\pi}{2}\right)+\frac{1}{3}\cos\left(3\omega_0 t-\frac{\pi}{2}\right)+\frac{1}{5}\cos\left(5\omega_0 t-\frac{\pi}{2}\right)+\cdots\right] \tag{2.7}$$

**Table 2.1** First terms of the series that forms the periodic function.

| **$4/\pi$ sen$\omega_0 t$** | **$4/3\pi$ sen$3\omega_0 t$** | **$4/5\pi$ sen$5\omega_0 t$** | ***t*** | ***f*(*t*)** |
|---|---|---|---|---|
| 0 | 0 | 0 | 0 | 0 |
| 0.39345 | 0.34335 | 0.25464 | $T/20$ | 0.99144 |
| 0.7438 | 0.403639 | 0 | $T/10$ | 1.15201 |
| 1.0300 | 0.13115 | −0.25464 | $3T/20$ | 0.906 |
| 1.2109 | −0.2494 | 0 | $2T/20$ | 0.9615 |
| 1.273236 | −0.424412 | 0.25464 | $5T/20$ | 1.1034 |
| 1.2109 | −0.2494 | 0 | $3T/10$ | 0.9614 |
| 1.0300 | 0.1311 | −0.254647 | $7T/20$ | 0.906 |
| 0.74838 | 0.4036 | 0 | $2T/5$ | 1.1519 |
| 0.39345 | 0.3433 | 0.254647 | $9T/20$ | 0.9914 |
| 0 | 0 | 0 | $T/2$ | 0 |

Fig. 2.5B represents the harmonic functions from the previous relation, the ones that, added point to point, as can be seen in Table 2.1, reproduce the square wave. So, the more terms the series has, the more accurate will be the representation of the original function.

In its exponential form, the Fourier series is defined as follows:

$$f(t) = \sum_{-\infty}^{\infty} c_n e^{jn\omega_0 t} \tag{2.8}$$

Where the absolute value of the coefficients $c_n$ is given by

$$|c_n| = \left|\frac{1}{2}\sqrt{a_n^2 + b_n^2}\right| \quad \text{with} \quad C_n = 2|c_n| \quad \text{and} \quad C_0 = c_0 = \frac{1}{2}a_0$$

The graphic representation of the coefficient amplitude $c_n$ with respect to the frequency $\omega$, that is, in the frequency domain, is known as the frequency spectrum or amplitude spectrum. Likewise, the phase spectrum is the representation in the frequency domain of the phase angle between signals, defined by

$$\phi_n = \tan^{-1}\left(-\frac{b_n}{a_n}\right) = -\theta_n \tag{2.9}$$

Fig. 2.6 presents the spectrum of frequency and phase of the square wave analyzed in the previous example.

In general, the frequency spectrum of a periodic function is a discrete spectrum obtained from the exponential form of the Fourier series. However, if the period of the function tends to infinity ($f(t) \rightarrow \infty$), the produced

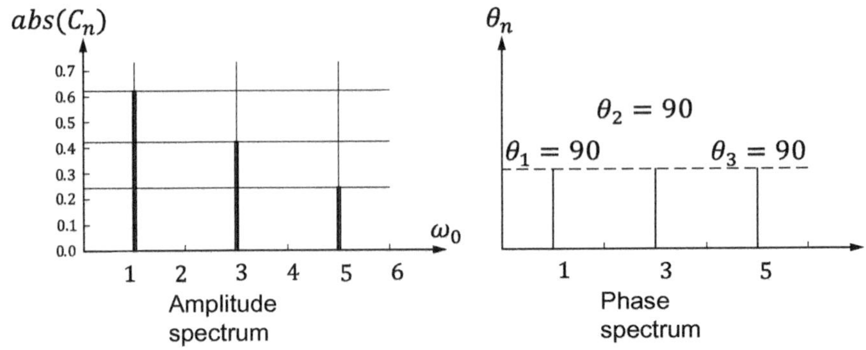


**Fig. 2.6** Frequency spectra and of square wave phase of Fig. 2.5.

spectrum is continuous and is defined by what is known as Fourier's integral or Fourier's transform. This transform is given by

$$F(\omega) = \int_{-\infty}^{\infty} f(t)e^{-j\omega t}dt \tag{2.10}$$

This expression defines the pass of a signal from the time domain to the frequency domain. Putting into a graph $|F(\omega)|$ with respect to the frequency $\omega$, the continuous spectrum or the magnitude spectrum of the function $f(t)$ is obtained. If passing a signal from the frequency domain to the time domain is required, it would have the inverse function of the Fourier transform defined as follows:

$$f(t) = \frac{1}{2\pi}\int_{-\infty}^{\infty} F(\omega)e^{j\omega t}d\omega \tag{2.11}$$

The two previous relations allow a signal to pass from the time domain to the frequency domain and vice versa.

## Fast Fourier transform

According to the relations and concepts outlined in the previous sections, an algorithm was developed that, by being programmed in a microprocessor, allows obtaining the frequency spectrum of a signal coming from a vibratory phenomenon. This signal is periodically sampled during a determined interval and by applying the so-called fast Fourier transform (FFT) algorithm to it, its components are obtained in the frequency domain.

The FFT is the basis of the spectral analysis and of the instruments generically known as spectra analyzers. Given that the calculations made by the FFT algorithm are based in the sampling of an analog signal, that is, in the time domain, care is needed in the sampling frequency, in the time interval to be sampled, in the resolution of the expected response, in the units to be used, and in the bandwidth of interest for analysis of the problem that is being studied.

The application of the FFT carries out the transformation from the time domain to the frequency domain in an optimal way if the dataset in the time domain contains $2^n$ elements, where $n = 1,2,3, \ldots$, taking into account that the signal processing in the microprocessor is based precisely on the binary system.

Associated with the discrete character of the signal to be handled by the FFT, the effects mentioned below are present.

*Aliasing*: The *aliasing* effect is associated with a *sampling error*. This error consists of the effect that two signals of different frequencies may look like both signals are the same if the sampling frequency is too low or if high frequencies are not filtered. That is, when the signal is being sampled in the time domain, an effect similar to the one produced by a stroboscope may occur, meaning that a high-frequency signal may be confused with a low-frequency signal. This would produce errors in getting the frequency spectrum. This effect is illustrated in Fig. 2.7, where it can be seen that if two signals of different frequencies are spotted at a low sampling frequency, they would produce a similar output signal.

As a general rule (Nyquist theorem sampling), it is recommended to sample a frequency of at least twice the highest frequency value present in the range of interest. In the spectra analyzers, the *aliasing* is eliminated by passing the signal to be sampled through filters that allow the passing of a certain range of frequencies to ensure that the maximum frequency of the range of interest is no greater than half the sampling frequency.

*Signal resolution*: The vibratory signal that varies with respect to time is *analog* and requires being converted into a *digital* one that could be handled numerically in a fast way. To do so, analog-digital converters (A/D) are used. These devices allow obtaining from an analog time domain signal a numerically coded signal ($N_1$, $N_2$, …, $N_i$) that, in turn, could be handled with a binary code, that is, with a combination of ones and zeros (Fig. 2.8). The use of microprocessors of greater or lesser capacity conditions the number of points the digitized signal has, that is, the maximum resolution of the sampled signal.

The capacity of the A/D convertor conditions the number of binary digits (*bits, **bin**ary dig**its***) available to register the information. An A/D converter of 8 bits of capacity will allow the use of $2^8 = 256$ intervals while one

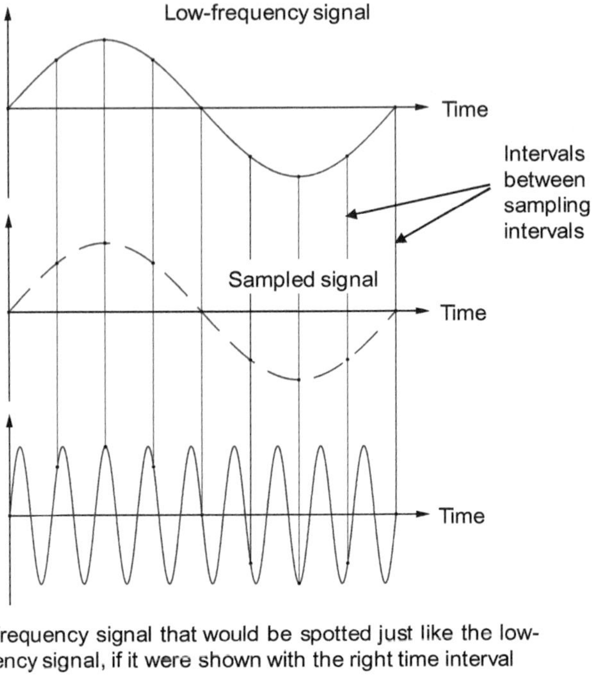


**Fig. 2.7** Aliasing effect.

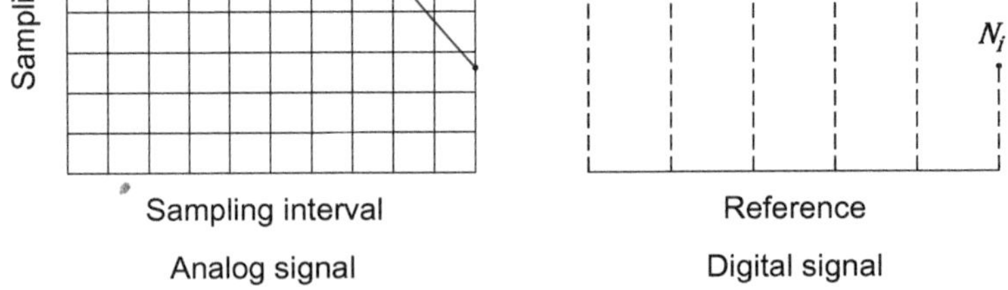


**Fig. 2.8** Conversion from an analog signal to a digital one.

of 12 bits will allow the use of $2^{12} = 4096$. So, the greater the interval number, the better will be the resolution of the signal to be analyzed.

*Bandwidth*: Having a predetermined number of points of the sampled analog signal also conditions the resolution of the frequency spectrum resulting from the FFT application. The *bandwidth* is the difference between the maximum frequency and the minimum one to sample. The *bandwidth* selection is then the range of frequencies in which the analog signal is to be analyzed and determines the quantity of information and details that could be observed in the spectrum.

Depending on the equipment available, the signal resolution on the frequency domain will vary. In an *X*-*Y* axes graph, the resolution is the number of intervals in the *X* axis (frequency) available for a proper observation of the spectrum under study. For a fixed number of resolution lines, the greater the selected bandwidth, the lesser will be the definition of the frequency spectrum peaks.

The bandwidth selection and the maximum frequency to be analyzed are important issues in predictive maintenance because one has to choose in order not to miss relevant signals when analyzing a specific problem. In rotating equipment, the bandwidth selection will depend basically on the maximum equipment rotation speed. Generally, in the case of a wide-band (see the "Application of the spectral analysis to the study of vibrations" section) spectrum analysis, the minimum frequency is equal to zero when selecting the bandwidth.

## Window

Considering that the signal to be analyzed is continuous, an interval has to be selected in which the FFT will be applied. This is accomplished by establishing a window in which the analog signal to be analyzed is represented. The establishment of the window consists of multiplying the sampled signal by a window function, which allows eliminating the discontinuity between the sampling periods in a way that makes the data displayed as zeros at the beginning and end of the considered interval. To illustrate this concept, Fig. 2.9A presents a periodic signal with a selected window. In this case, the sampling periods will reproduce the signal with no error. In respect to a transitory signal such as the one shown in Fig. 2.9B, this window only allows covering part of the signal, so that the sampling of these intervals would look like the one shown in Fig. 2.9C. The discontinuity gets eliminated by applying a window function (as shown in Fig. 2.9D), leaving the signal to be analyzed by the FFT as illustrated in Fig. 2.9E.

In transitory signals, not applying a window function when using the FFT would make a spurious component appear in the frequency spectrum

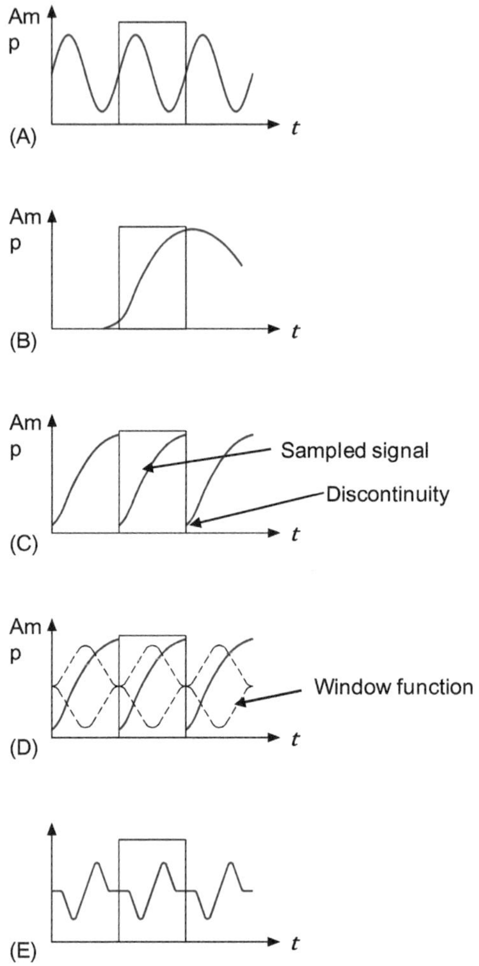


**Fig. 2.9** Effect of windowed signals. (A) Sampling of a periodic signal, (B) sampling of a transitory signal, (C) the FFT varies the signal as a discontinued signal, (D) application of a window function, and (E) signal that will analyze the FFT.

related to the frequency of the discontinuities of the analyzed signal. There are diverse window functions such as the rectangular, the HAMMING, the Kaiser-Bessel, and the HANNING, which are the most commonly used. Fig. 2.10 represents qualitatively how a signal would appear if, for example, the Hamming or Kaiser windows were applied.

The window function to be applied will depend on the type of problem as well as the type of analyzer available.

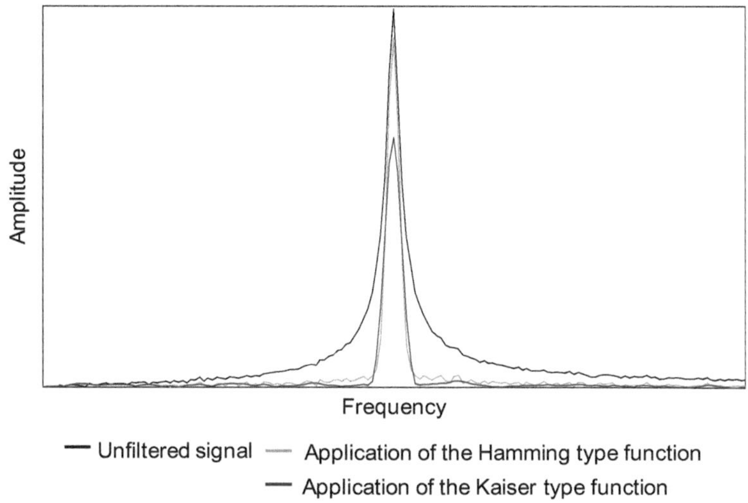


**Fig. 2.10** Application of window functions to a signal.

## Application of the spectral analysis to the study of vibrations

Most of the equipment used in industrial plants can be considered mechanical systems that produce noise and vibrations. This is the reason why vibration analysis is the most frequently used technique for handling predictive maintenance programs. Based on the theoretical foundations exposed in previous sections, electronic equipment and computer programs have been developed industrially that allow the spectral analysis of vibrations to be carried out in a fast and relatively direct way. This relative ease of equipment use is precisely why it is necessary to know the fundamentals of the Fourier analysis to make correct interpretations of what was reported by the different equipment. Once the fundamentals of the spectrum analyzers are established, this section presents some of the capacities of this equipment for carrying out both the recording and analysis of these signals.

*Broadband and narrowband*: The measurement and analysis of vibrations produced by machinery help determine the operating conditions in a way that any change in the characteristics of the vibration signal can be detected and associated with the vibration of machine components. Vibration signal analysis is carried out in a certain range of frequencies $\Delta\omega$, that is, in a certain bandwidth, which is related to the value of the maximum frequency to be considered and taking into account the resolution of the available spectrum analyzer. Depending on the number of components of the vibration signal measured that can be identified in the frequency domain, analysis can be made in a *broad bandwidth (broadband)* or in a narrow bandwidth (*narrowband)*. Although this terminology is somewhat ambiguous, in the context of predictive maintenance, a *narrowband* analysis is carried out when the signal under study has two well-defined peaks. This means that there are components of the signal of significant value just around the frequency at which the peak occurs. In general, this type of analysis is carried out when information about the condition of a specific machine component is required. The *broadband* analysis is carried out when there is a signal with components whose amplitude is similar to that of the central signal of the band and is applied when information is required about a machine's overall state or a set of machine components.

In spectrum analyzers, there are available filters that allow the pass of a percentage of the tuned central frequency and filters that make possible the pass of signals in a fixed bandwidth, making it possible to select the type of analysis, either *broadband* or *narrowband*. Fig. 2.11 presents the difference of the bandwidth obtained by applying a constant magnitude width, or the one

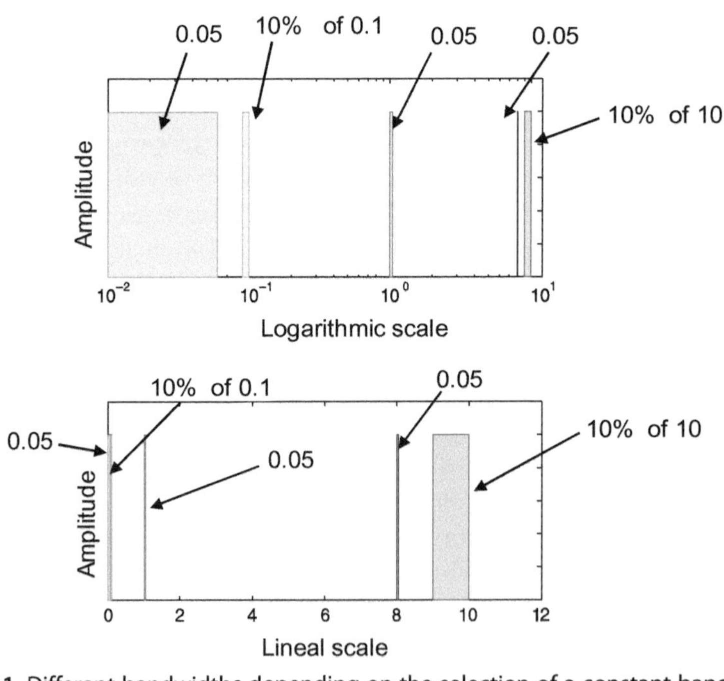


**Fig. 2.11** Different bandwidths depending on the selection of a constant band or one constant percentage band.

obtained if a constant percentage of the central frequency is applied, in both a linear and a logarithmic scale.

The selection of the bandwidth depends on the application and availability of the equipment for analysis. Special care should be taken on the band selection as there may exist important differences regarding the obtained information from the resulting spectra, as can be seen in Fig. 2.12.

*Tendencies of vibration.* In Fig. 2.13A, a mechanical system is shown where measurements at several points take place. An analysis of the tendency of vibration, selecting a broadband, would provide information about the overall machine condition at the measurement point, obtaining a frequency spectrum such as the one shown in Fig. 2.13B. This spectrum, defined by using the root mean square value RMS, would provide a measurement of the total energy present in the selected bandwidth. The registration of the variation tendency of the RMS value concerning the energy of the system (see the "Energy content of a signal" section), as can be seen in Fig. 2.13C, would provide a measurement of the global evolution of the machine condition, but it would not give more information concerning the state of some

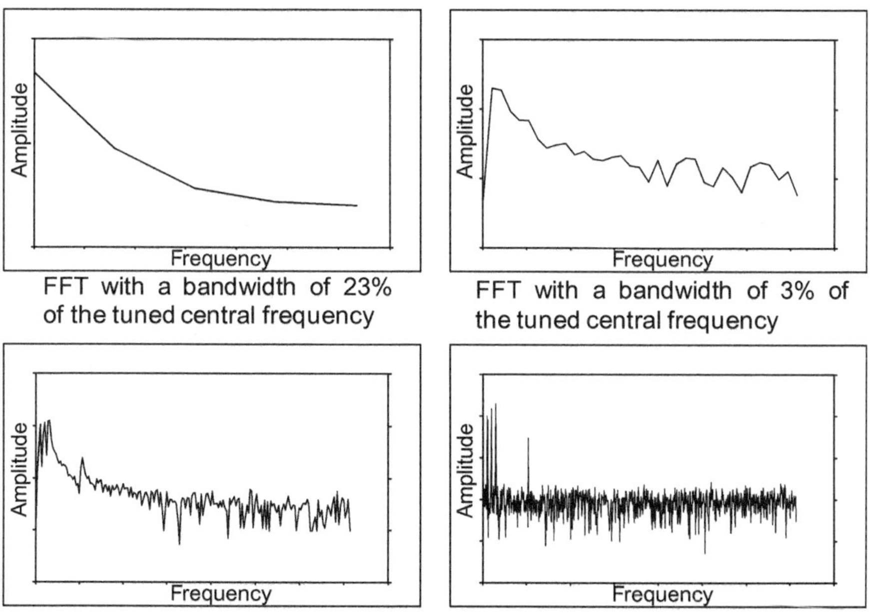


FFT with a bandwidth of 23% of the tuned central frequency

FFT with a bandwidth of 3% of the tuned central frequency

FFT with a bandwidth of 0.5% of the tuned central frequency

FFT with a bandwidth of 0.14% of the tuned central frequency

**Fig. 2.12** Effect of the bandwidth variation in the resolution of the signal.

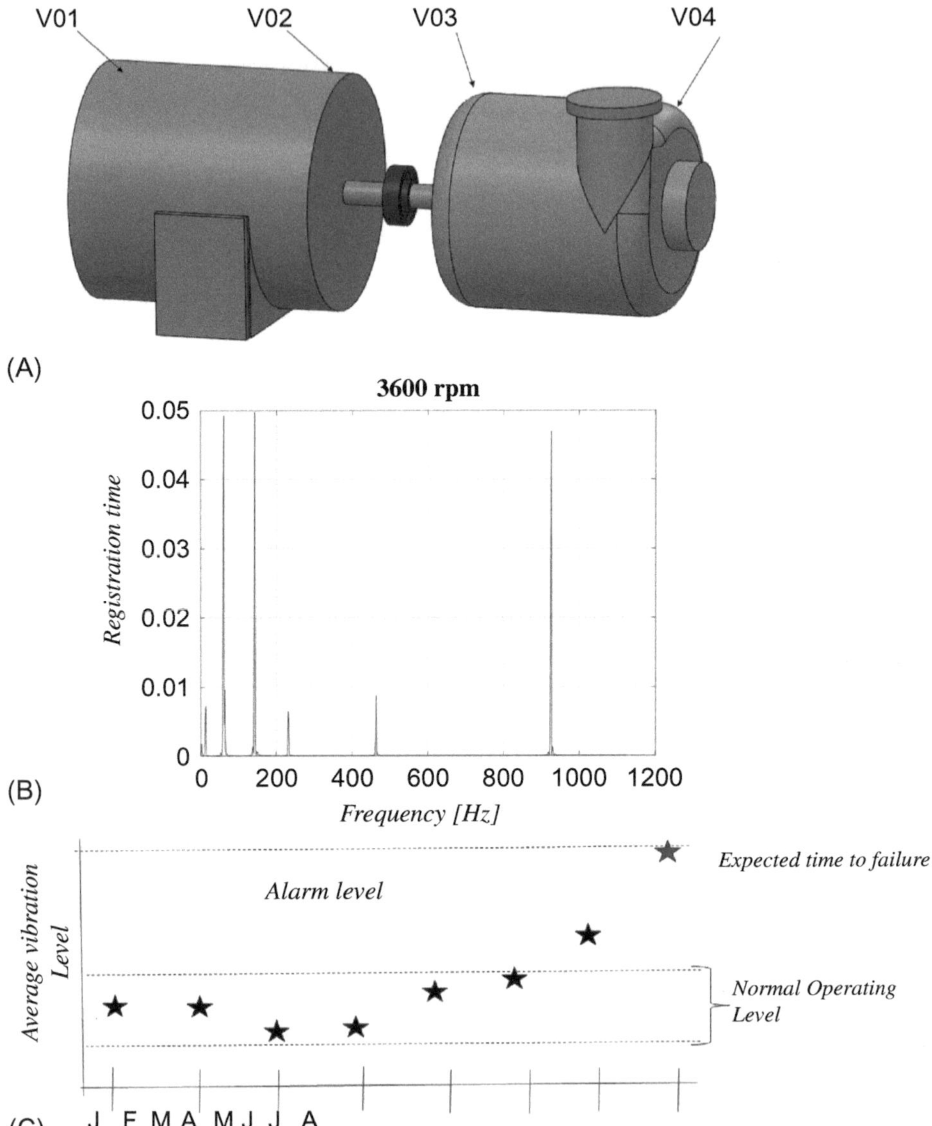


**Fig. 2.13** (A) Measurement of the vibration in several points of a mechanical system. (B) Typical spectrum of vibration for a narrowband analysis. (C) Example of a variation tendency of the root mean square of the vibrations of the ball bearing to be monitored (V02), by giving a monitoring periodicity (see Chapter Seven).

of its particular components. When the latter is required, a narrowband analysis is carried out.

## Vibration signatures

The signal obtained in the frequency domain of a certain machine or mechanical system, in normal operating conditions, is known as the vibration signature. Based on this signal, the behavior of the machine can be compared when carrying out new analysis, allowing us to continue monitoring the evolution of its various components through its lifespan in operating conditions.

The main technique used for continuing the evolution process of the vibrations in a mechanical system is based on the recording of the variation tendencies of the vibration amplitude compared with its vibration signature. This recording can be carried out by selecting a band, either broad or narrow, and observing the variation of the characteristic spectrum, which produces a certain component of a machine in known operating conditions. In Fig. 2.14A, a frequency spectrum in normal operating conditions is shown, and Fig. 2.14B presents the spectrum for a fault condition.

The tendency of the vibration amplitude variation of a determined system or equipment can be recorded from the vibration signature; these are known as cascades, which are consecutive records of spaced spectra for a certain period of time. In Fig. 2.15, a cascade spectra is shown in which can be seen the amplitude and frequency changes of a vibration signal with time.

## Signal averaging

By applying the FFT, the averaging of the signal to be analyzed allows a better definition of the spectrum. In most analyzers, there are options to average linearly, exponentially, or in a synchronized way with another signal.

In the lineal average, the instant spectra are summed one by one and the total is divided between the number of considered spectra. In exponential averaging, more importance is given to the latest recorded spectra than to the initial ones. The synchronized averaging allows linking the start of an analysis to an operation point or to a specific moment of the equipment operation under study. In Fig. 2.16, the effect of averaging a signal is shown.

## Measurement units

The selection of measurement units when registering machine vibrations is an important part of the spectral analysis because the correct result interpretation

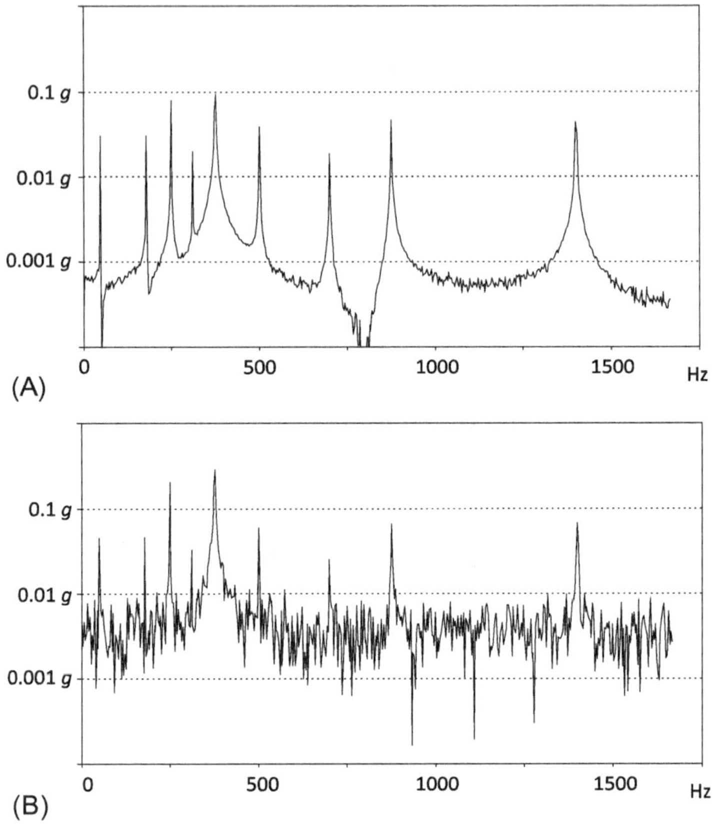


**Fig. 2.14** (A) Vibration signature of a machine and (B) registered/recorded signal under fault conditions.

depends on it. The units of the variables to be considered must be clearly stated, either by their vibration amplitude, their speed, or their acceleration. Special care should be taken with the measurement system being used because in the industry, it is still common to use the English system (foot, pound, second); however, the application of the International System of Units (meter, kilogram, second) is becoming more popular.

In order to present the variations of multiple variables in a unified way, in spectral analysis it is common to measure in decibels (dB). These units allow expressing in a summarized way the changes present while monitoring the behavior of different machines involved in a predictive maintenance program. Once the normal level is set up, a direct relationship between the new measurements determines the critical operating conditions.

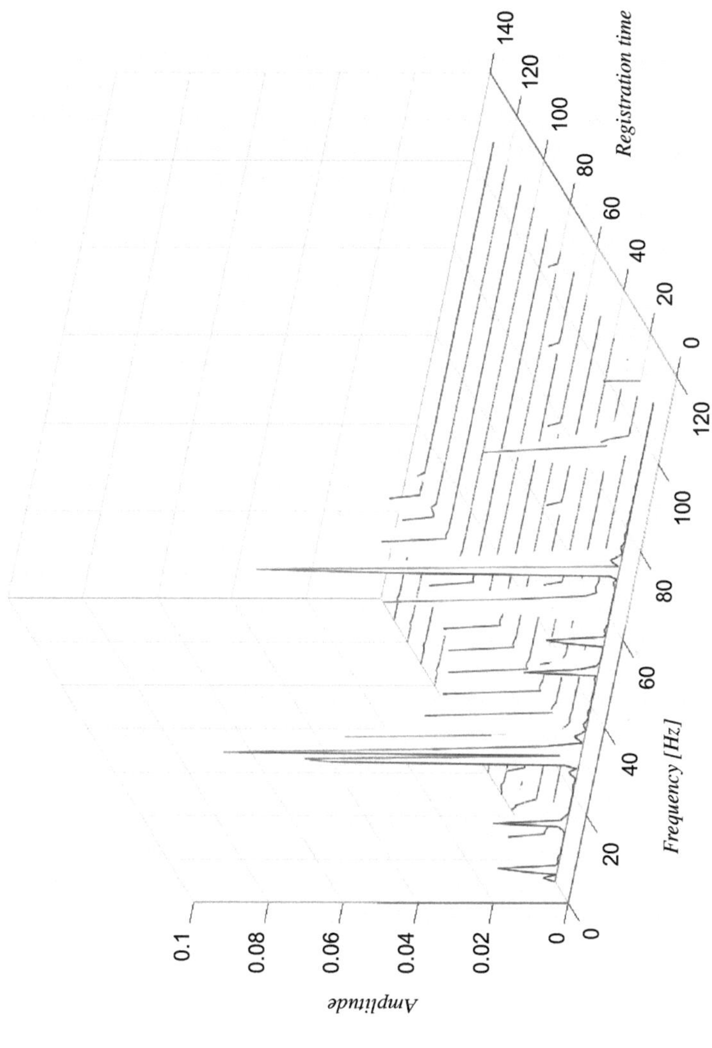


**Fig. 2.15** Cascade record of a frequency spectrum.

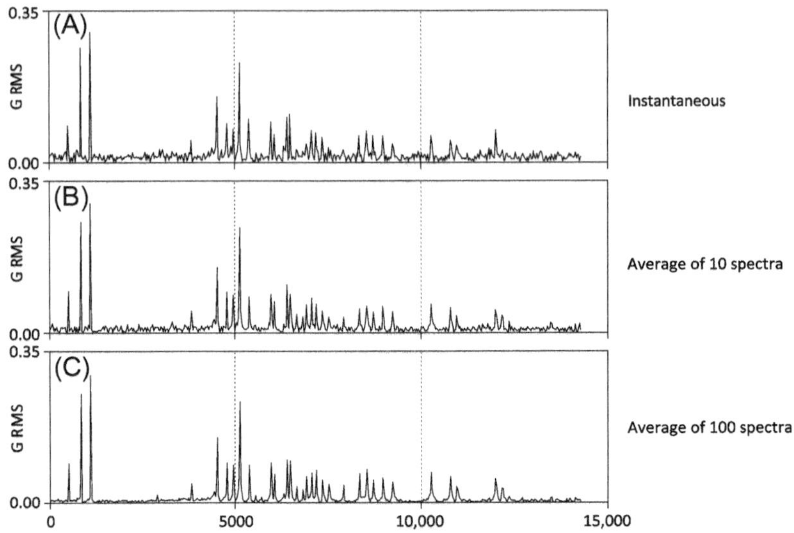


**Fig. 2.16** Effect of averaging the measurement of a signal. (A) Instantaneous measurement, (B) averaged 10 times, and (C) averaged 100 times.

The expression of the decibel in the predictive maintenance area is

$$\mathrm{dB} = 20\ \log_{10}(N_1/N_2)$$

In this formula, $N_1$ is the level of final vibration and $N_2$ the level of initial or reference vibration.

Decibels express a relation between two numbers, so it is possible to establish a level of reference that indicates the level 0 dB. In vibration analysis, it has been internationally agreed that the reference for speed is $1\times10^{-9}$ m/s (RMS) and for acceleration is $1\times10^{-6}$ m/s$^2$ (RMS). For example, for a speed level of $10^{-5}$ cm/s, there would be a level of +40 dB. The plus sign (+) indicates a level of increase while the minus sign (−) indicates a decrease or attenuation.

Although other references are set up, the dB always involves relative changes of level, so that a change from 10 to 1 between two quantities will always be −20 dB and one of 1 (initial value) to 10 (final value) will be +20 dB. In Table 2.2, a series of equivalencies for several relative changes of level are shown for a certain signal, expressed in dB.

Once the spectral analysis has been carried out, selecting the units for the deployment of results is required. The proper selection of these units facilitates the identification of causes and effects of the analyzed vibrations. In this

**Table 2.2** Values of some relations of change expressed in dB.

| Relative change from $N_1$ to $N_2$ | dB |
|---|---|
| 1 to 1 (no change) | 0 |
| 1.4125 a 1 | 3 |
| 2 a 1 | 6 |
| 3.16 a 1 | 10 |
| 10 a 1 | 20 |
| 31.6 a 1 | 30 |
| 100 a 1 | 40 |
| 1000 a 1 | 60 |
| 10,000 a 1 | 80 |
| 100,000 a 1 | 100 |

sense, linear and logarithmic scales are available for both the amplitude and frequency. Fig. 2.17 presents the difference that a signal would have with the displayed frequency in linear and logarithmic scales while Fig. 2.18 shows how the amplitude changes if it is specified in one or another scale.

When displaying the results of a spectral analysis on the analyzer screen, it is also important to consider the signal resolution so that a detailed interpretation can be carried out and, if necessary, an amplification can be made. In this way, the frequency spectrum can be observed with more accuracy, as can be seen in Fig. 2.19.

Note that the scale of logarithmic amplitude must be selected when the levels of change between one signal to another are not so significant because this scale magnifies the small amplitude values. Otherwise, there is a risk (that happens regularly) of not detecting the vibration of low energy that is generated in the first stages of a mechanical failure.

The type of vertical scale (displacement, speed, or acceleration) is selected according to the signal to monitor. It is recommended to select *displacement* units to monitor problems of low frequency and great amplitude; *speed* units for all types of vibration signals; and *acceleration* units to detect high-frequency and low-amplitude signals.

## Energy content of a signal

In vibration analysis, there are situations from which it is required to know the energy content of a signal in a certain range of frequencies. Most spectra analyzers have this ease-of-use feature, which, conceptually, in the time domain is represented as follows:

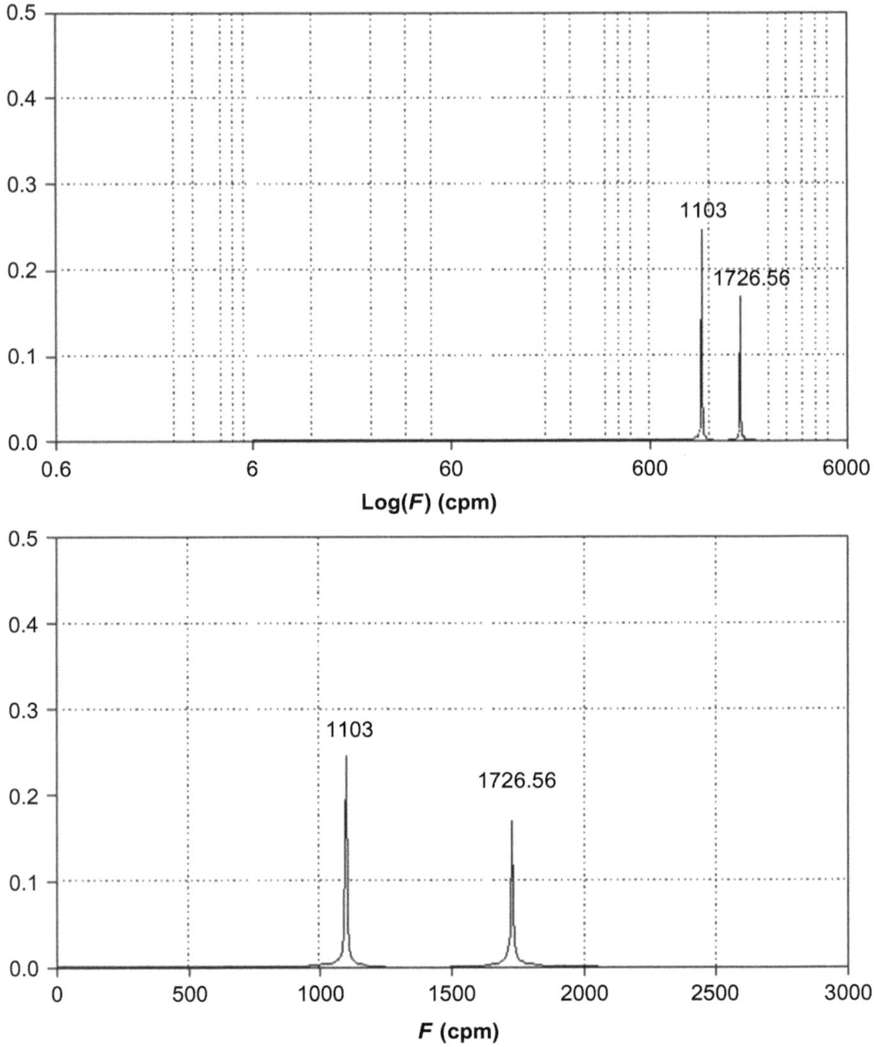


**Fig. 2.17** Representation of a signal, varying the scale of frequency.

$$E = \int_{-\infty}^{\infty} |f(t)|^2 dt \tag{2.12}$$

In the frequency domain, the energy content of the signal is called energy spectrum or spectral energy density, and it is defined by the expression:

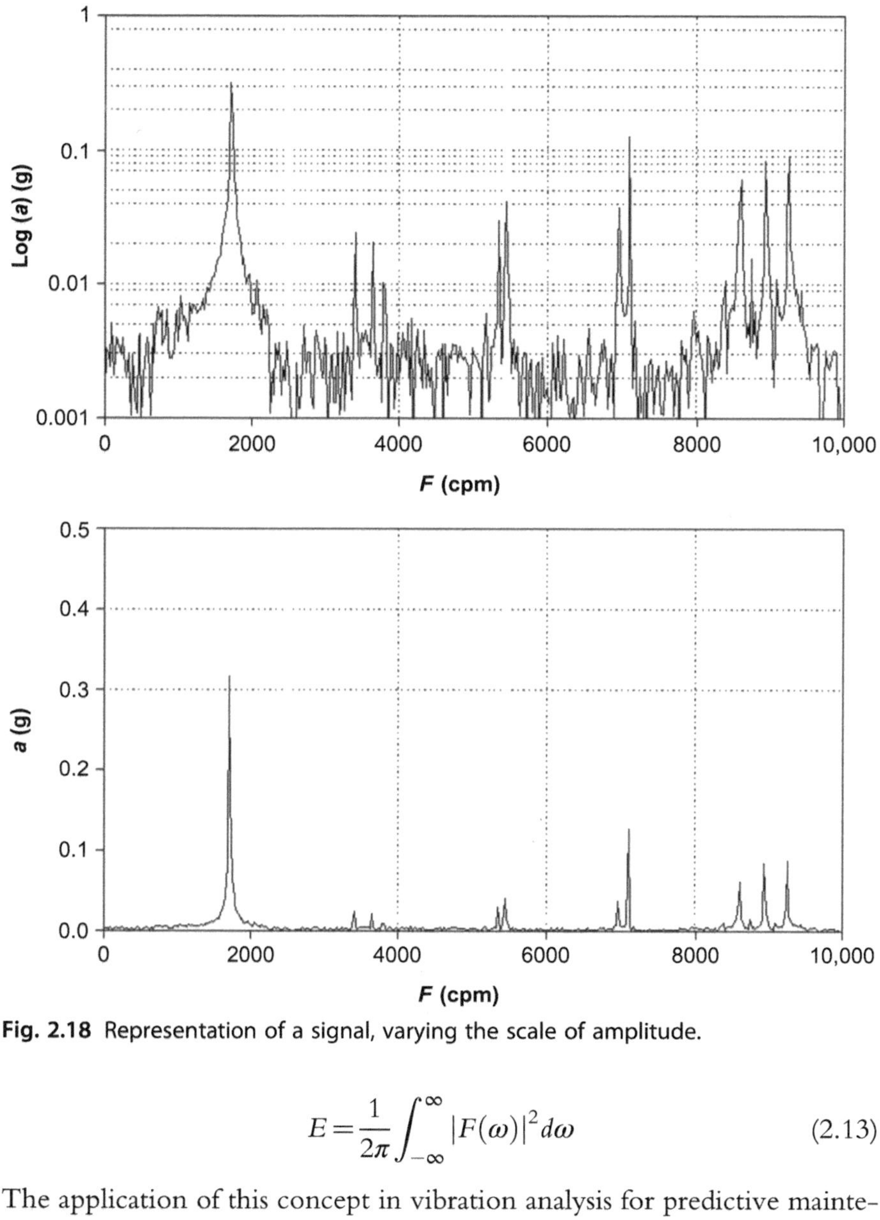


**Fig. 2.18** Representation of a signal, varying the scale of amplitude.

$$E = \frac{1}{2\pi}\int_{-\infty}^{\infty} |F(\omega)|^2 d\omega \tag{2.13}$$

The application of this concept in vibration analysis for predictive maintenance is very useful because it allows knowing the energy associated with a certain effect. The spectral energy density can be obtained for a certain range of frequencies, that is, for a certain bandwidth, $\omega_{\text{initial}} < \omega < \omega_{\text{final}}$ associated with the effect under study. This also would allow identifying the causes that produce the vibration.

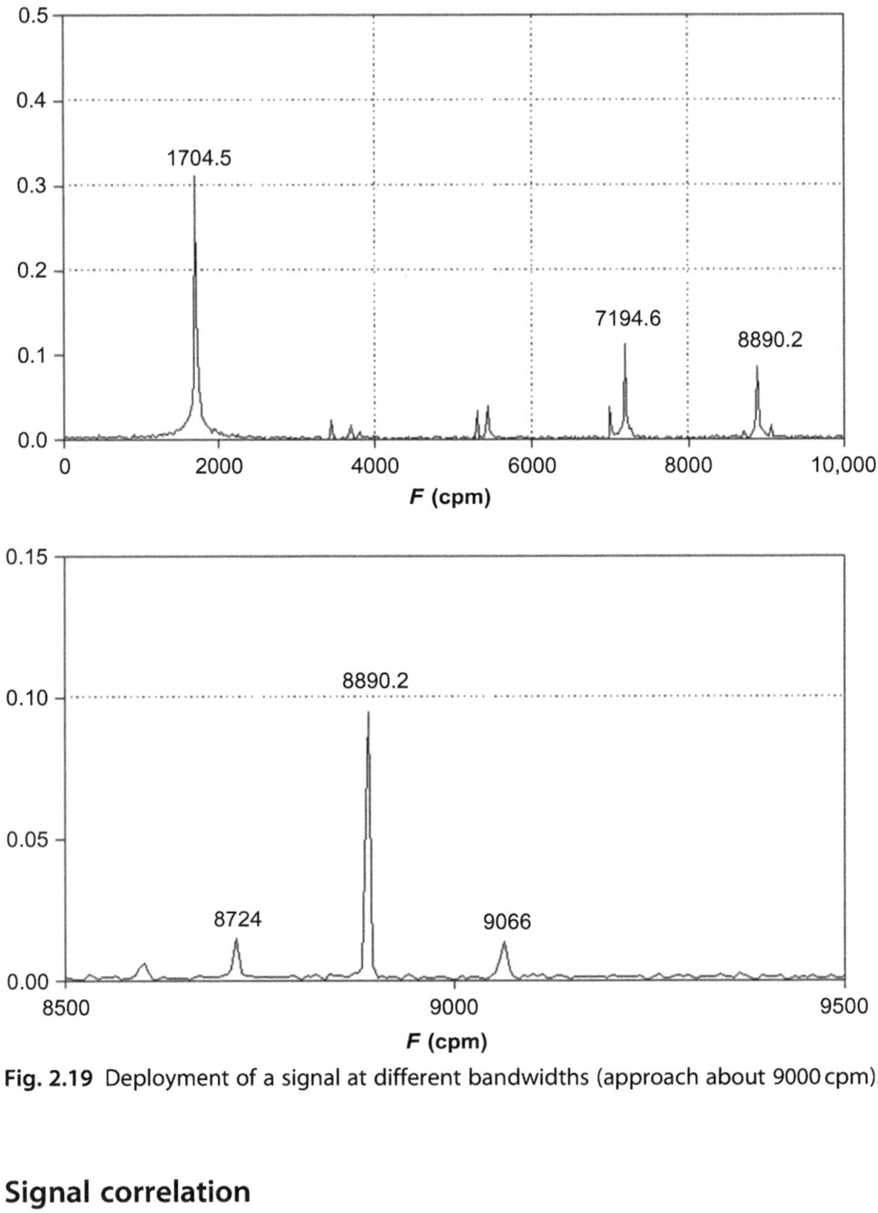


**Fig. 2.19** Deployment of a signal at different bandwidths (approach about 9000 cpm).

## Signal correlation

Another useful concept in spectral analysis applied to predictive maintenance is the so-called *correlation function*. This function gives a measure of the similarity between two functions. That is, if two functions are similar with respect to their frequency and amplitude, the correlation function will highlight the common frequency components to both functions.

The mathematical expression that defines the correlation function is

$$R_{12}(\tau) = \int_{-\infty}^{\infty} f_1(t) f_2(t-\tau)\, dt \tag{2.14}$$

This expression represents the correlation between functions $f_1(t)$ and $f_2(t)$, occurring at a different time. The case in which $f_1(t)$ and $f_2(t)$ are equal is known as autocorrelation. The application of this function in predictive maintenance allows the characteristics of a vibratory signal to be identified, even if the signal has a lot of noise, and also checks the compatibility of measurements carried out on the same machine at different times.

Consider the square wave of amplitude a, shown in Fig. 2.20A. Applying to this wave the autocorrelation function at time $t_0$ and considering the signal random phase shift $(t-\tau)$ between 0 and $T$, the correlation function shown in Fig. 2.20E is obtained. For a phase shift $\delta = T/4$, the autocorrelation value is minimum (Fig. 2.20E, point $A$). This means that the degree of similarity or interdependency between the signals (a) and (b) is minimum at that time. If the autocorrelation function is applied when $\delta = T/2$

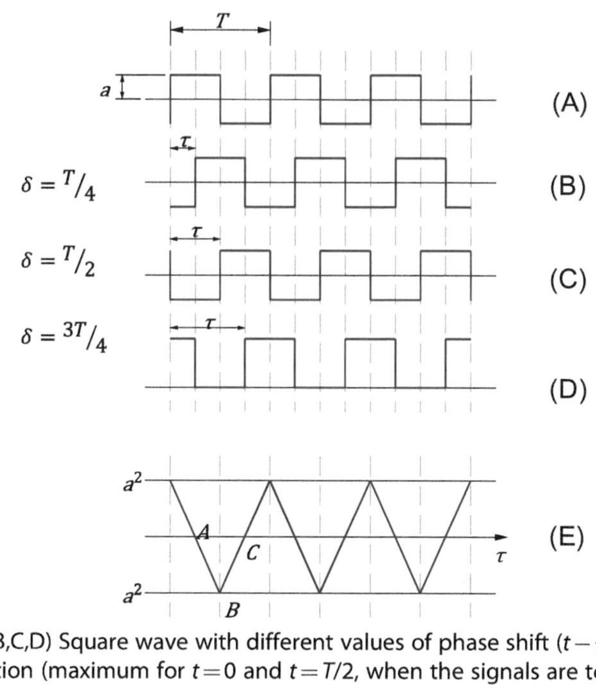


**Fig. 2.20** (A,B,C,D) Square wave with different values of phase shift $(t-\tau)$. (E) Autocorrelation function (maximum for $t=0$ and $t=T/2$, when the signals are totally in or out of phase).

(Fig. 2.20C), the autocorrelation has a maximum value, as can be seen in Fig. 2.20E, point B.

The autocorrelation of the square wave when $\delta = 3T/4$ (Fig. 2.20D) turns out to be, again, a minimum value, as can be seen in the point $C$ of Fig. 2.20E. The autocorrelation of the shown square wave, to any value of $\delta$, is the triangular function of amplitude $a^2$ shown in Fig. 2.20E. In this figure, it can be seen that the correlation between $f_1(t)$ and $f_2(t)$ has a maximum positive when the signals are in phase, and a maximum negative when they are out of phase. Applying the autocorrelation to a random signal coming from a machine vibration allows verifying the reliability of the measurements by checking the amplitude of the autocorrelation function that should keep its maximum amplitude at a constant level.

The correlation of functions $f_1(t)$ and $f_2(t)$ when coming from different sources is called cross-correlation. Application of this function in vibrations analysis for predictive maintenance allows identifying the relation and influence that two signals coming from different components of a machine have on each other. For example, if in a machine there is a signal produced by some ball bearing and another measured in the casing of the machine, the correlation function could be applied to both signals to determine the importance of the ball bearing component in the vibration of the casing.

Fig. 2.21 illustrates this concept. In this figure, two signals are shown (Fig. 2.21A and B), one square wave of amplitude $b$ ($f_1(t)$) and a frequency of 41.1 Hz, and a sine wave with an amplitude $c$ ($f_2(t)$) and a frequency of 13.7 Hz. When applying the cross-correlation to these functions, the correlation function shown in Fig. 2.22A is obtained. The frequency spectrum of this correlation function is shown in Fig. 2.22B. From this spectrum, the relation between the frequencies of both signals and the harmonics of the square signal can be seen.

Applying the cross-correlation concept to a predictive maintenance case study would allow identifying in the cross-correlation frequency spectrum possible resonances of a component of the machine under study, excited by the vibration of another component of the same machine or from other nearby equipment.

## Noise

In the analysis of machinery vibrations, there are occasionally recorded signals that do not have any information concerning the problem under study. These are signals whose source could be either a bad connection

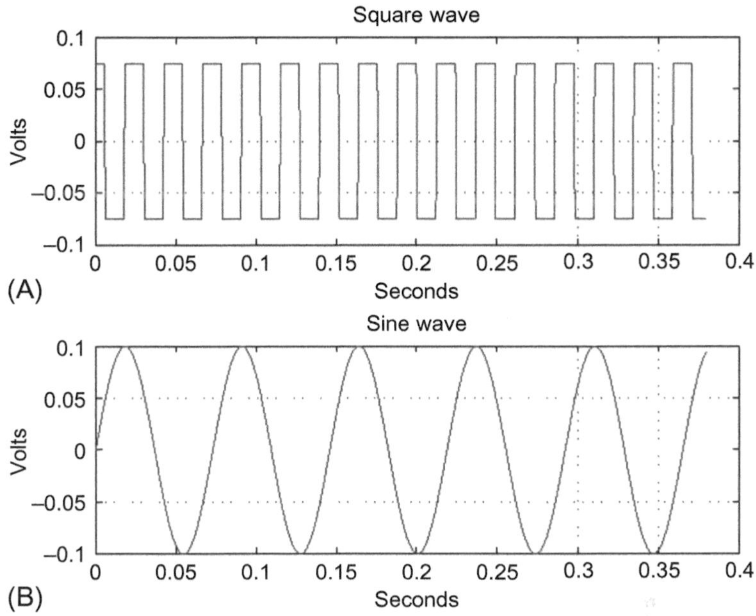


**Fig. 2.21** Signals to which the cross-correlation will be applied: (A) square wave and (B) sine wave.

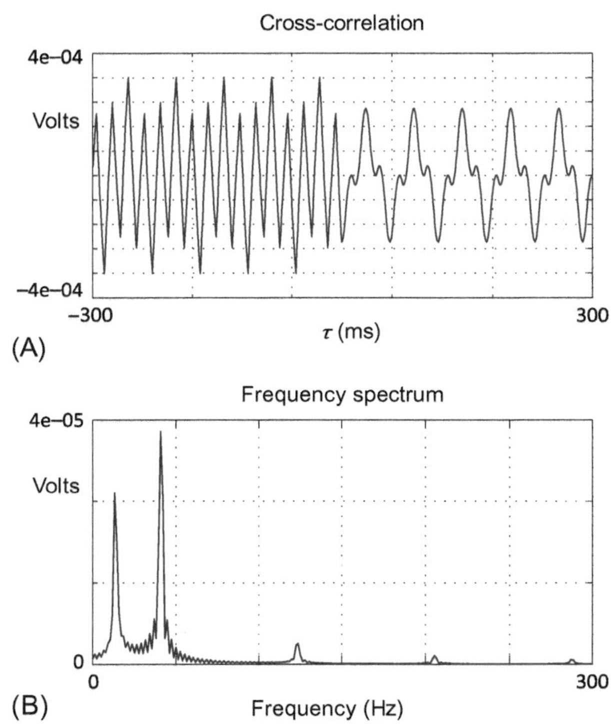


**Fig. 2.22** (A) Cross-correlation waveform and its (B) frequency spectrum.

of the equipment used or electromagnetic interference. These types of signals are called noise. A special kind of noise inherent to the operation of the instruments is the so-called white noise. Even though this noise is of very low amplitude (of the microvolt order), it must be considered when interpreting the result signals. It can be detected if the average correlation function is close to zero, which would mean that there is no relation among the components of the noise signal.

CHAPTER THREE

# Instruments

## Introduction

Vibration analysis is the most common technique of predictive maintenance programs in the industry because most of the equipment of industrial plants is mechanical. This technique is based on the measurement of both the noise and the level of vibrations to determine the operating conditions of the different components of a mechanical system. Setting up a predictive maintenance program will depend, to a large extent, on the instruments selected, the ease of operation, and especially the ability to interpret the results.

A typical machinery monitoring system is composed of a spectra analyzer, a computer, a data-handling program, and the transducers. In general, the monitoring system must allow both the recording and handling of data in an automated form in such a way that it should be possible to obtain accurate and reliable operational tendencies of the different machine components. Of the four components of the aforementioned system, there is much diversity of commercial products, and it is almost impossible to analyze all possible variants to form a monitoring system. However, by knowing the operating principles of the system, it will be possible to select and use the most appropriate equipment according to the operating conditions in each case.

This chapter presents both the operating principles of the piezoelectric accelerometers and the spectra analyzers, describing the characteristics of the data-handling program. Practically, any computer equipped with a hard drive may be used in a predictive maintenance program, and its performance will depend on the characteristics of the data-handling program.

## Data-handling program

The program applicable to handling collected data taken from the predictive maintenance system is normally integrated into the equipment. Transmitting data from one program to another might seem simple.

*Mechanical Vibrations and Condition Monitoring*
https://doi.org/10.1016/B978-0-12-819796-7.00003-2

However, it gets more complex because generally these programs do not have normalized input or output formats. For this reason, special care should be taken when selecting the equipment for a vibration monitoring program because any extra equipment needed for improving the monitoring program must be compatible with that initially acquired. The vibration monitoring system must include features such as the capability to organize a database, provide tendencies of measurements, set alarm levels, and automatically generate reports.

Program operation: In general, these programs are organized on menus. As a fundamental characteristic of these kind of programs, it is necessary that they have a data protection system to avoid deleting the collected data by accident. At the same time, the program should have facilities for editing and setting up data presentation that are easily available to the user.

In selecting the operating program for the acquisition system, it is important to make sure it is capable of acquiring and processing the data collected so that vibration signal time variation can be visualized. The higher the data storage capacity, the more expensive the predictive maintenance system will be. Therefore, it will be necessary to get the optimum point between the computer cost and the program one as well as the response speed and the amount of data to be stored.

Another required characteristic of the computer program is that it must allow the automatic generation of reports and be able to combine as many variables as possible. Another useful feature of the program to be selected is the possibility of listing all machine vibration levels that have exceeded previously set values. The program must also allow setting up levels considered for alarms for different machine components under study, and automatically generating signal tendencies supported by stored data. Taking into account the available capacity of modern computers, it is therefore feasible to get a good capacity of data handling, allowing the user to compare the historical vibration tendencies of machinery.

## Spectra analyzer

These types of instruments allow transferring data from the time domain to the frequency domain; that is, they have fast Fourier Transform programming (FTT) integrated. Even though the use of powerful microprocessors in both the computer and the spectra analyzer make vibration analysis straightforward, a solid knowledge of the principles of vibration theory and analyzer operating principles is needed to get maximum knowledge from the data collected.

## General description of a spectra analyzer

Fig. 3.1 presents the block diagram of a spectra analyzer, which uses the FFT as a base algorithm for the transformation of the analog input signal from the time domain to the frequency domain.

Despite the great variety of spectra analyzers that have been developed, all consist of basically an analog input block, an output processor block, and a controller block. The latter provides the data sent through a high-speed serial port from and to a microprocessor, which controls the data deployment on a screen. This microprocessor can be a personal external computer in which the data-handling system of predictive maintenance also runs.

Input signals are sent from the analog input block to the preamplifier, where an overload test is carried out, before being sent to an analog filter that prevents sampling error or aliasing. The filtered signal is sampled at high frequencies through a digital-analog converter (A/D). The spectra analyzers normally used for predictive maintenance carry out the sampling at

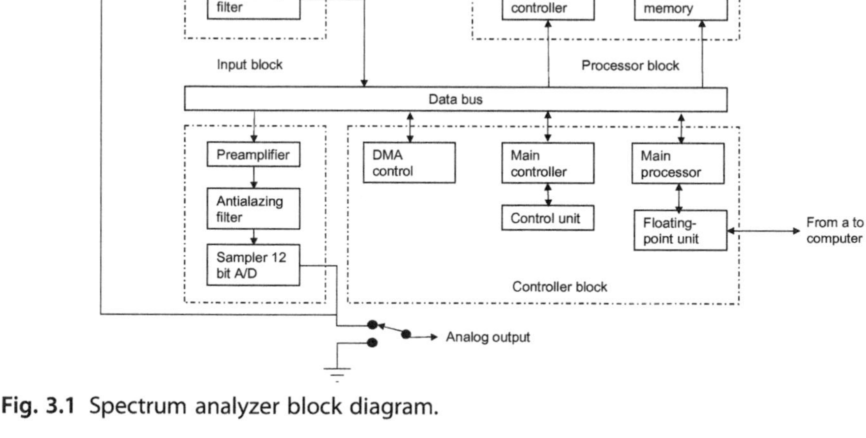


**Fig. 3.1** Spectrum analyzer block diagram.

frequencies of about 50 kHz with 12-bit A/D converters. Once this sampling finishes, there is a digital signal available. This signal, another overload test, and a new digital filtering process are carried out to avoid aliasing. At this stage, the sampling frequency changes according to the selected bandwidth. This frequency is kept at around 2.5 times the value of the maximum frequency of the defined bandwidth.

Some analyzers have the option of zooming, which allows a higher resolution to observe details around a certain frequency. During the entire conditioning process of the analog signal and its digitalization, the spectra analyzers have to ensure that all sampling errors (aliasing) have been attenuated. This attenuation is usually expressed in decibels (dB).

Once it has been digitized, the signal is sent to the memory through a system called direct memory access, or DMA. This system ensures fast and accurate data transfer through what in electronics is known as a databus.

In the microprocessor, the FFT algorithm and the windowing functions are programmed. Also, the voltage levels of the digitalized signals are verified to detect a possible saturation of voltage levels handled by the microprocessor. Normally, the signal processing signal time is of the order of 30 ms for an arrangement of 1024 sampled points.

Subsequently, in the output processor block, a microprocessor with a mathematical coprocessor converts the results of the FFT application, which are whole numbers, in an arrangement of floating point numbers that can be sent for graphing and/or displaying on the computer screen. In this same block, everything related to the limits of the *X* and *Y* axis of the graph and the type of scales to be used is programmed, taking into account the type of device to be used for the spectrum display.

Some analyzers have an analog output block that allows the digitalized signals to be handled again as analog signals.

Operating parameters of the analyzers: To efficiently process the input data to the spectra analyzer, it is necessary to consider the relation that exists between each of its operating parameters. That is, the bandwidth (*ab*), the number of resolution lines, (*lr*), the sampling time (*tm*), the sampling frequency (*fm*), the amount of points to be sampled (*cp*), the interval between two consecutive resolution lines ($\Delta f$), and the interval of time between two points of the sampled analog signal ($\Delta t$).

Then, the relations among these parameters are

$fm = n \cdot ab$ where $n \geq 2$ (according to the sampling theorem of Nyquist)

$$tm = lr/ab, \Delta t = 1/fm, cp = tm/\Delta t, \text{ and } \Delta f = 1/tm.$$

## Transducers

The main transducers used for the detection of vibrations include those for displacement, speed, and acceleration. However, the accelerometers are the most frequently used for their accuracy, measuring range, ease of assembly, and cost. In addition, it is relatively simple to numerically integrate the acceleration signal and obtain the speed and displacement.

## Piezoelectric accelerometer

The piezoelectric accelerometer is a device that produces a proportional electrical signal to the acceleration. It consists of a piezoelectric crystal fixed to a base by means of a mass. The base is fixed to the vibrating element in such a way that the whole set of the accelerometer moves with the vibrating element. As the accelerometer moves, the mass in its interior produces a pressure on the piezoelectric crystal, generating an electrical charge ($q$) that produces a difference of potential $E$, as shown in Fig. 3.2. Considering that the force is proportional to the mass by the acceleration ($F=ma$) and that the load is proportional to the acceleration, the accelerometer delivers an electrical signal directly proportional to the applied force.

The load produced on the walls of the piezoelectric crystal is given by $q=CE$, where $q$ is the load given in coulombs, $C$ is the capacitance whose units are the farads, and $E$ is the voltage. In the accelerometers, the load normally generated is very small, so it is measured in pico-coulombs ($10^{-12}$ coulombs).

The piezoelectric accelerometer is the most common transducer for measuring vibrations and can be represented as a single-degree-of-freedom

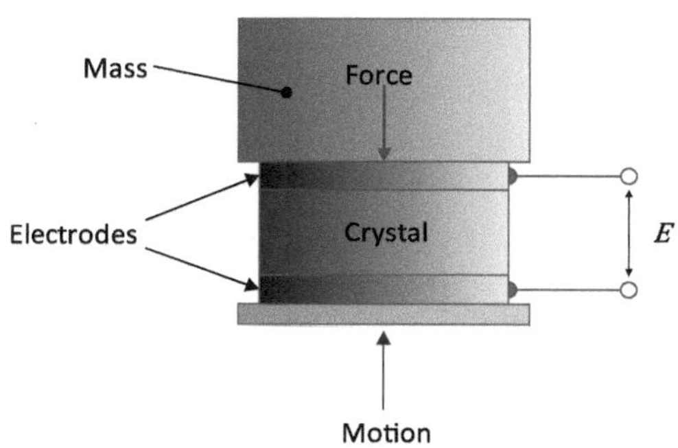


**Fig. 3.2** Operating principle of a piezoelectric accelerometer.

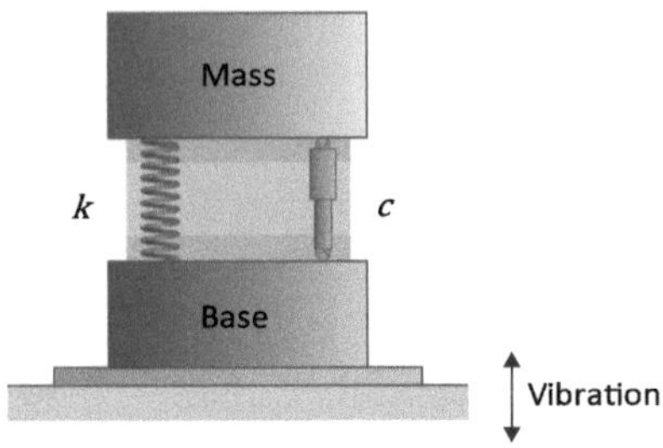


**Fig. 3.3** Equivalent model of an accelerometer.

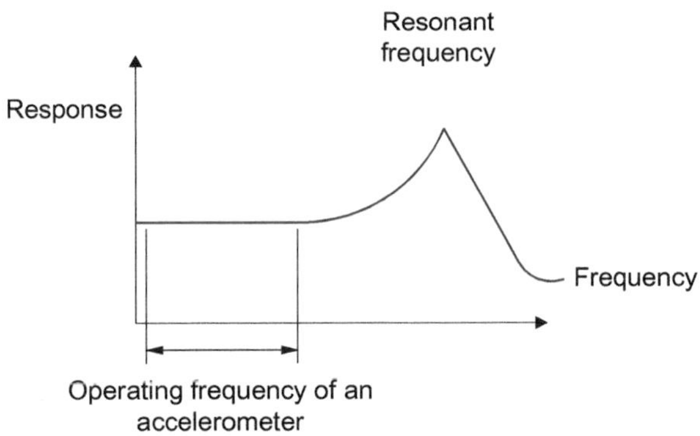


**Fig. 3.4** Typical response curve of an accelerometer.

system mass-damper spring, as shown in Fig. 3.3. The response of this system in the time domain was described in the first chapter. Depending on the value of the natural frequency, the response of the system in the frequency domain can be seen in Fig. 3.4. That is, if the system is designed to have a very high natural frequency, the accelerometer will have a near constant range of response. In this range, a linear correspondence is obtained between the input motion and the output electrical signal.

Because the operating principle of the accelerometers is associated with an inertial force, they are called seismic accelerometers. However, this work refers to the aforementioned device simply as an accelerometer.

Considering the accelerometer electrical circuit parameters, they are classified according to their output impedance. There are low and high impedance accelerometers, as described below.

High impedance accelerometers: These types of accelerometers are called high impedance because the sensor element is directly connected to the output of the circuit in such a way that the output impedance is of the order of $10^{12}\ \Omega$. These types of accelerometers are also called load

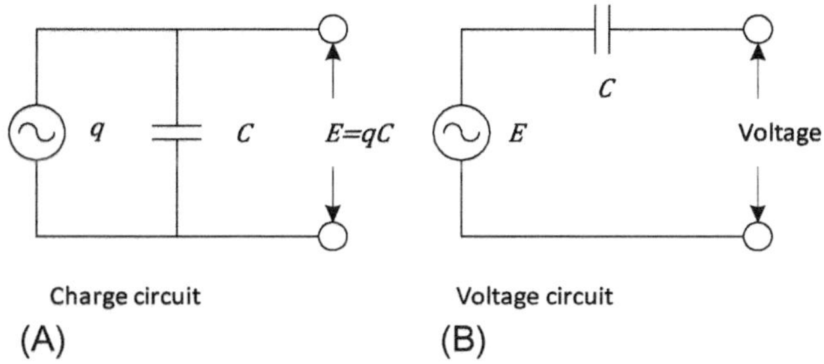


**Fig. 3.5** Equivalent circuit of a piezoelectric crystal: (A) of load and (B) of voltage.

accelerometers because the applied load is proportional to the force applied. Normally, they are used in high-temperature environments that do not allow integrating any type of electronic circuit within the transducer itself. Considering the voltage generated between the piezoelectric crystal walls, the equivalent circuit of these types of accelerometers can be considered a load generator (Fig. 3.5A) or a voltage generator (Fig. 3.5B).

These types of sensors work as a source of high capacitive impedance. This characteristic poses special problems when measuring the load and voltage for sending them to some distant monitoring point to the point of measurement. This operation requires the use of either a charge amplifier or a voltage amplifier.

Load amplifier: It is an amplifier of a very high gain with negative capacitive feedback that keeps the input voltage close to zero. With this feedback, the load input produces a voltage in the feedback capacitor. This voltage is equal to the value of the load input divided by the feedback capacitance. The output voltage is practically equal to the voltage in the feedback capacitor because the input voltage is neglected compared with the input one. So, the feedback capacitor determines the relation between the output voltage and the input load, as can be seen in Fig. 3.6.

The relationship among the input load ($q_e$), the output voltage ($E_s$), and the feedback capacitor ($C_r$) can be deduced by observing Fig. 3.6, which suggests the following:

$$E_s = \frac{E_e(C_t + C_c)}{C_r} = \frac{q}{C_r} \tag{3.1}$$

Here, $q$ represents the load generated by the transducer, represented as $q_e$.

The major advantage of the load amplifiers is that they precisely perceive the charge, which is a parameter of the transducer that is not affected by the cable length. This allows the calibration of the charge according to the

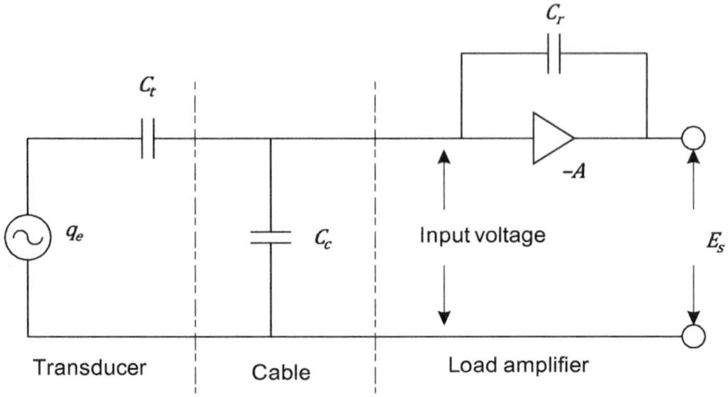


$A$ = Amplifier gain

$C_r$ = Capacitance of the feedback capacitor

$C_c$ = Cable gain

$C_t$ = Capacitance transducer

**Fig. 3.6** Simplified circuit of an accelerometer with a load amplifier.

sensibility of the transducer (expressed in pico coulombs per acceleration unit). However, being a high gain operational amplifier whose amplification point covers the entire cable length, it is particularly sensitive to the electrical noise generated in the cable. That is, the level of the transducer output signal does not vary with the cable length, but this signal gets contaminated by the level of noise generated by the cable. This makes essential the use of a high-quality coaxial cable with a good shield to ensure low levels of noise. It is also required to have the cable fixed for preventing the generation of low frequency noise produced by cable movement. Using filters to prevent the passing of low frequencies would limit the passing of information as well, making their use not recommended.

The noise generated by the cable movement makes uncertain impact measurements using load amplifiers due to the cable flexion and the sudden movement during impact. The signal generated by the cable is often higher than the one generated by the transducer, so there are repeatability problems of the measurement.

The voltage amplifier considers the transducer as a voltage generator in series with a capacitor, as can be seen in Fig. 3.7. In this way, the capacitance of the connector cable is connected in parallel, resulting in a voltage divider that is independent from the frequency but not from the cable length.

The residual noise in a system that uses a load amplifier increases as the cable length or the input capacitance rises while the level of the resulting

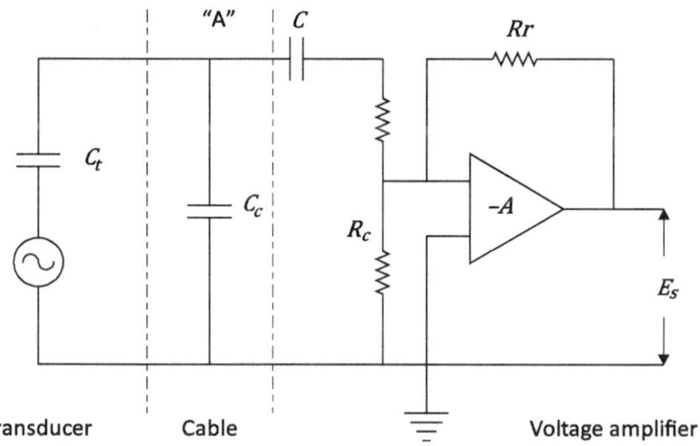


**Fig. 3.7** Simplified circuit of an accelerometer with a voltage amplifier.

signal is not affected. In contrast, in a system that operates with a voltage amplifier and uses a cable whose length or capacitance is greater, the level of the input signal decreases, leaving the noise level relatively constant. So, in both load and voltage systems, the relation between the level of the input signal and the level noise deteriorates as the cable length increases, though for different reasons.

Load amplifiers with high impedance transducers are the most used combination in industry due to their output sensitivity and independence of the cable length and its capacitance. However, it is necessary to take into account that a low noise cable needs to be used as well as to know that the circuit is very sensitive to moisture, pollution, and environmental noise.

Low impedance accelerometers: In low impedance accelerometers, the electronic circuit required to detect the voltages generated by the piezoelectric crystal is integrated as part of the sensor. In this way, the transformation from high to low impedance is made in the measuring point, and only low impedance signals are transmitted from the sensor to the data collector. This design uses a microcircuit integrated into the accelerometer, resistant enough to impacts. It only requires two cables from which the power is supplied and the signal is received. For these types of accelerometers, the sensibility is not affected by the cable length and does not need external signal conditioners.

In Fig. 3.8, a typical circuit for these types of accelerometers is shown. It can be seen that the voltage registered in the resistance $R$ is detected by a field effect transistor (*MOSFET*). This transistor is energized with a direct current of constant level and is polarized by a voltage of reference. As the

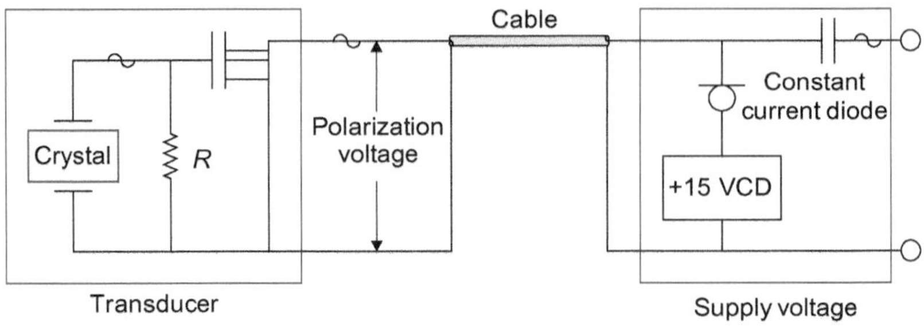


**Fig. 3.8** Simplified circuit of a low impedance accelerometer.

piezoelectric crystal generates voltage, the impedance of the field effect transistor increases or decreases linearly with this voltage. The circuit is coupled to the power supply through a capacitor. The minimum response frequency of the system is controlled by the value of the resistance $R$ and the internal capacitance of the piezoelectric crystal.

These types of circuits have gains of about 0.95, with the output voltage slightly lower than that of the amplifier input. This decrement between the input and output voltages means a gain in the circuit current. Likewise, they keep their linearity close to 80% of the level of polarization voltage. The output impedances of these circuits are about 500–1000 Ω and the polarization voltage varies between 5 and 8 V. Output impedances of up to 50 Ω can be obtained with higher polarization and supply voltage, allowing it to have greater sensitivity.

In low impedance accelerometers, additional circuits are integrated for signal processing, such as filters that allow only the passing of some frequencies or which compensate the operation according to the environmental temperature. In these types of accelerometers, outputs up to $1\,V/g$ can be obtained, responding to frequencies lower than 1 Hz as well as noise levels of $10^{-6}\,g$ ($g$, gravity acceleration).

The main advantages of low impedance accelerometers are that they do not depend on load amplifiers, have very wide operation ranges, and have very high sensitivity and little sensitivity variation related to the cable length.

Construction of the piezoelectric accelerometer: In Fig. 3.9, a cross-section of a piezoelectric accelerometer is shown. In this figure, it can be seen that the crystal is contained between the so-called seismic mass and the base, which is rigidly connected to the accelerometer capsule. This design has been found to be resistant enough to mechanical loads and changes in temperature.

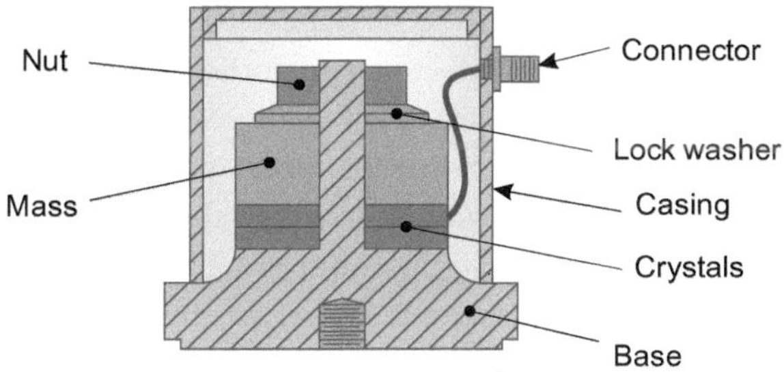


**Fig. 3.9** Cross-section of a piezoelectric accelerometer.

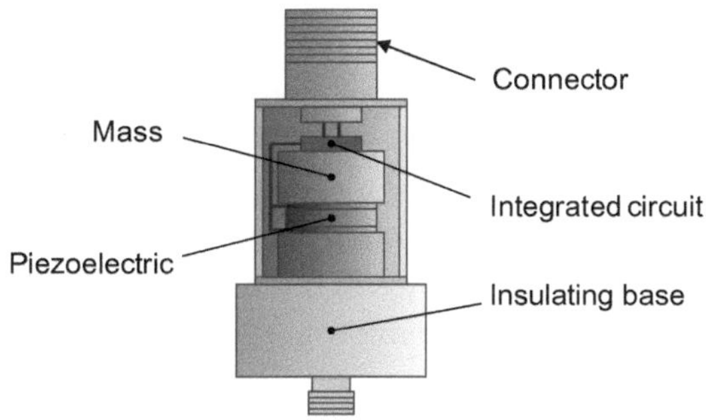


**Fig. 3.10** Cross-section of an electrically isolated accelerometer.

Depending on the operating conditions of the accelerometer, its electrical isolation needs to be considered. So, there are electronically isolated accelerometers such as the one shown in Fig. 3.10.

Main sources of noise: The term "noise" is generally accepted as any undesired signal present in the accelerometer output. The main sources of electrical noise in accelerometers are the crystals and their associated microcircuits, the noise coming from the surface on which they are mounted, and the radiofrequency coming from equipment around the accelerometer. Some types of piezoelectric crystals produce low-intensity spurious signals, making it necessary that the microcircuit associated with the accelerometer has the appropriate filters to keep the noise threshold between 10 and 20 μV.

To avoid the noise coming from the surface where the accelerometer is mounted, it is recommended to electrically ground that surface. Also, to eliminate the ground loops, it is common practice to install an insulating nut between the base and the insulating material such as nylon, cellophane, Bakelite, or plastic.

Temperature effects on piezoelectric materials: The temperature changes as well as the high temperatures affect to a greater or lesser extent some properties of the piezoelectric materials, such as the capacitance, the resistivity, and the piezoelectric constant. Some materials present an increase in the capacitance as the temperature increases, others show a decrease, and some behave differently to distinct temperatures. These variations produce changes in the charge and voltage sensibility of the accelerometer. Under extreme conditions, the material may lose entirely its piezoelectric properties.

The variation of piezoelectric material properties with temperature has applications in heat-transfer problems, but for their use in accelerometers, this temperature property variation represents an additional variable to be considered when analyzing vibration measurements in variable-temperature environments.

Mounting a piezoelectric accelerometer: Mounting the accelerometer to the mechanical element to which measurements are going to be made is a key issue for the correct performance of the accelerometer. Variations in the mounting stiffness might produce important changes in the signal obtained from the accelerometer. Depending on the application, the fastening can be made through a threaded mount, gluing the accelerometer, using a magnetic base, or holding it with a support rod.

Fastening the accelerometer through a threaded is the method that ensures greater stiffness. It consists of screwing the accelerometer by means of a stud bolt directly to the vibrating surface. With this mounting method, it is required to ensure the perpendicularity between the surface and the axial shaft of the accelerometer. In this case, to avoid deformations in the base of the accelerometer, special care must be taken with the tightening torque because this may affect the piezoelectric element.

The gluing method is limited to measurements of relatively low frequencies because the adhesives used do not ensure an absolute stiffness and may suffer deterioration with temperature. However, this fastening practice to fix an accelerometer is very common for the versatility in choosing the measuring point.

Fastening the accelerometer by means of a magnetic base is widely used in industry due to the ease of locating the transducer, even under difficult access conditions. However, this type of union can produce signals with a high noise content, which in some cases can decrease the linear range of the accelerometer by up to 50%.

The fastening method through a support rod is the least recommended. Measured frequencies may vary up to a third part with respect to those measured with a threaded fastening. This clamping method is limited to low-amplitude and low-frequency acceleration measurements and it is only used to make comparative measurements or in places subjected to high temperatures.

Fig. 3.11 shows a typical comparative response of an accelerometer depending on the fastening method used.

Parameters for selecting accelerometers: The most important parameters to specify for accelerometer selection are presented below. For each parameter, typical values and applications are presented that are related to the predictive maintenance of machinery.

Sensibility: The sensibility of an accelerometer is defined by the relation between the level of output voltage and the measured acceleration; it is normally expressed as $mV/g$, where $g$ is the acceleration of gravity. Therefore, an accelerometer will be more or less sensible if it generates more or fewer millivolts per each $9.81\,m/s^2$ of acceleration. Sensitivity values are given the specifying temperature and frequency at which they were measured, also indicating a tolerance. As an example, a typical sensibility for an accelerometer would be $100\,mV/g \pm 5\%$.

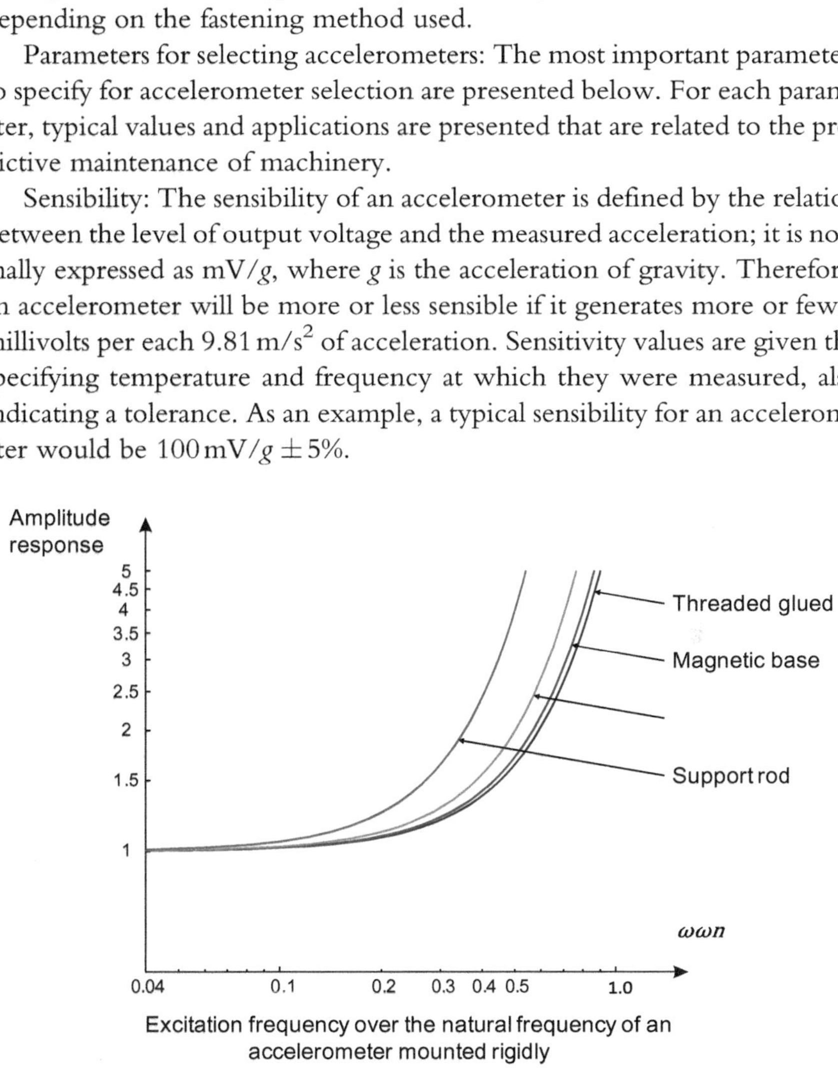


**Fig. 3.11** Variation of an accelerometer response depending on the fastening type.

Transversal sensitivity: It is the sensitivity an accelerometer presents in response to perpendicular accelerations to its axial axis. Due to machining imperfections in cutting or polishing the piezoelectric material, the geometric axis rarely coincides exactly with the highest sensibility axis of the piezoelectric material. This fact makes the transducer sensible to tangential accelerations. Transversal sensibility is expressed as a percentage of the accelerometer's main axis sensibility. A typical value of this parameter is 5%.

Frequency response: It is the range of frequencies in which the accelerometer response is linear within a certain tolerance. Generally, the sensibility of the accelerometer is specified to that presented to 100 Hz for a constant acceleration of $1g$ or $10g$. A typical frequency response range for an accelerometer would be from 3 Hz to 10 kHz, with a tolerance of ±5%. In some cases, it is possible to make measurements with variations beyond the specified percentage of tolerance if the results are adjusted according to the accelerometer response curve provided by the manufacturer.

Amplitude response: This parameter of the accelerometers refers to the maximum acceleration amplitude that it can register, specifying also the range in which its response is linear within certain tolerance. Normally, acceleration amplitude is specified with the maximum peak value that can be registered by the transducer. A typical value for this response in amplitude would be $\pm 80g$ (peak value), with a linearity of ±2% up to $\pm 64g$ (peak value).

Impact resistance: This parameter indicates the maximum acceleration that the accelerometer can withstand without damage. It is expressed in levels of $g$. A typical value would be $\pm 5000g$ (peak).

Resonance frequency: This quantity indicates the value of the frequency at which the accelerometer vibrates to its natural frequency, so its measurements are no longer valid. This frequency should be as far as possible from the measurement range. Normally, it is of the order of 2.5 times the maximum frequency of the accelerometer application. A typical value for this parameter related to mechanical vibration measurements is ≥25 kHz.

Noise level: This parameter indicates the value of the acceleration signal that would be present at all times during the operation of the transducer. A typical value for this parameter is $0.0002g$. As an example, for a sensibility value of $100\,mV/g$, there would be a signal of 0.02 mV, affecting the actual measurement.

Range and sensitivity temperature coefficient: These parameters indicate the temperature values where the accelerometer can operate as well as the percentage of variation that would have an output signal according to its working temperature. The typical operation range is between −50°C and 120°C, and the coefficient of sensitivity could even be expressed for two temperature ranges, for example, 0.03%/°C between −50°C and 40°C, and −0.05%/°C between 40°C and 120°C.

In addition to the parameters of operation mentioned above, the following parameters should also be taken into account: voltage and current requirements, output impedance, type of electrical insulation, type of connector, fastening device, seal type, materials of the capsule and its base, and the tightening torque. It is very important to consider that the mass of the accelerometer should not modify the dynamic response of the machine to be monitored.

## Velocity transducers

Velocity transducers are electromechanical sensors designed to directly measure vibratory movement. They are required to have a measuring element in motion, unlike the accelerometers that detect the applied exerted force upon a piezoelectric crystal rigidly attached to their base.

In many applications, there is the tendency to use accelerometers instead of velocity transducers due to the need for a periodic recalibration (typically, every 6 months). Accelerometers do not require such verification and may also detect vibrations on a wider frequency range. If an accelerometer is used, its output signal can be integrated to get the motion velocity or displacement. However, velocity transducers still continue to have application because their construction is relatively simple and their cost is lower compared to accelerometers. Velocity transducers have a low output impedance and do not require external energy sources to produce a signal, which can be transmitted using two wire conventional cables. Also, no major loss of sensitivity or pollution caused by noise coming from the cable is present. It is very common to find these types of transducers as an integral part of the monitoring equipment of all kinds of rotating machinery, especially in turbomachinery.

Operating principle of velocity transducers: The main components of velocity transducers are a permanent magnet, springs, a coil, and a casing. Fig. 3.12 presents a cross-section of a velocity transducer, where it can be seen how the coil is suspended by springs attached to the casing. The magnet is fixed to the casing, which has an air gap in such a way that the magnetic field lines are cut perpendicularly by the coil. When cutting the field lines, a

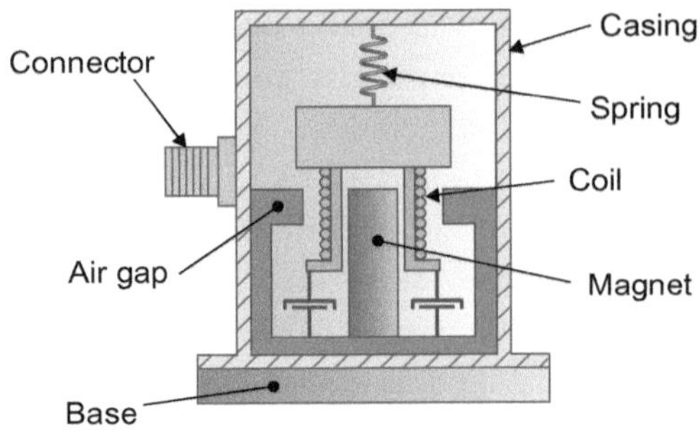


**Fig. 3.12** Diagram of the components of a speed transducer (cross-section).

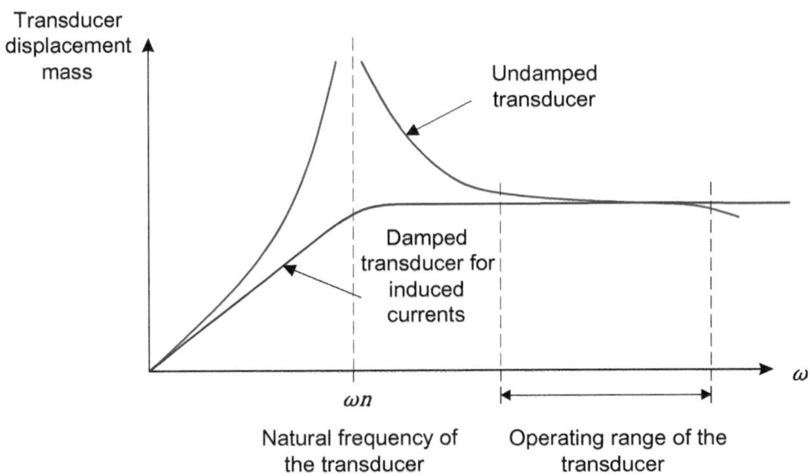


**Fig. 3.13** Operating range of a speed transducer.

potential difference is generated between the terminals of the coil, voltage that is directly proportional to the casing velocity with respect to the coil.

If the transducer natural frequency (spring-mass system) is lower than the frequency of the velocity variation, as can be seen in Fig. 3.13, the coil will remain practically stationary so the generated voltage will be proportional to the absolute casing velocity.

Most of the velocity transducers have a form of dampening. The most common one is based on induced currents, although viscous or electrical damping can also be found. The purpose of introducing a dampening is to limit the amplitudes of the mass (coil or magnet) when the transducer vibrates at frequencies close to its natural frequency (as can be seen in Fig. 3.13).

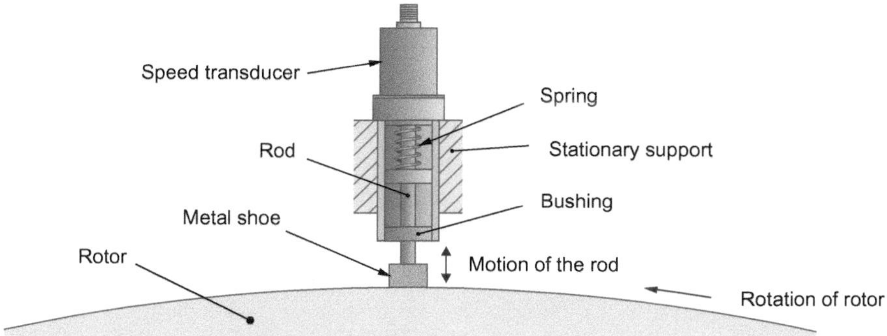


**Fig. 3.14** Moving coil velocity transducer.

Moving coil type velocity transducers: This type of transducer has a coil that moves along a magnetic field, giving a signal proportional to the velocity applied. In the case of a rotor, a guide in one of the ball bearings of the rotor and a spring ensure that contact is not lost. As can be seen in Fig. 3.14, the rotation speed is proportional to the vertical motion of the rod. That is, the velocity to be measured is applied to the rod. As the rod moves along the magnetic field produced by the coil, a voltage is generated that is proportional to the velocity to be measured.

Parameters to specify a velocity transducer: The main parameters are listed below. Units and typical values for applications in the predictive maintenance of machinery are indicated between brackets.

- Sensibility (from 34.9 to 53.15 mV/cm/s).
- Frequency response (from 10 Hz to 3 kHz).
- Temperature range (from −40°C to 280°C).
- Natural frequency (from 5 to 17 Hz).
- Mounting shaft (vertical or horizontal).
- Maximum acceleration (50*g*).
- Range of mass displacement (from 2.54 to 10 mm, peak-to-peak values).
- Output impedance (around 500 Ω).

## Eddy current displacement transducers

There are a great variety of operating principles for noncontact transducers. These transducers use capacitive, optical, and ultrasonic principles. There are variable reluctance types and those based on induced currents (eddy currents or Foucault currents). In this book, only the ones of eddy currents (induced currents) are described, as they are the most commonly used in programs of predictive maintenance of machinery.

The displacement transducers of the induced current type generate an electrical signal proportional to the clearance between the transducer and the vibrating surface, without these elements being in contact.

General description: The elements that form a displacement transducer of the induced current type are the probe, the extension cable, and the oscillator-demodulator. Fig. 3.15 presents the general arrangement of this transducer.

The probe: It is a small threaded steel body that has in one of its ends a spiral coil. The coil is protected by an encapsulation that is normally made of epoxy resin, ceramic, or fiberglass. This protective cover has a thickness of approximately 0.25 mm. Care should be taken with this cover because it can be easily destroyed if the probe is allowed to rub against some moving surface.

Extension cable: It is used to connect the probe to the oscillator-demodulator. The connection is made through a capacitive inductive resonant circuit, from which the probe is the inductive part while the cable is the capacitive element of the circuit. As the capacitance of the cable varies according to its length, it is important to use only the cables and probes for which an oscillator-demodulator has been calibrated. If the length of the extension cable is altered or if a probe is used with a connection cable

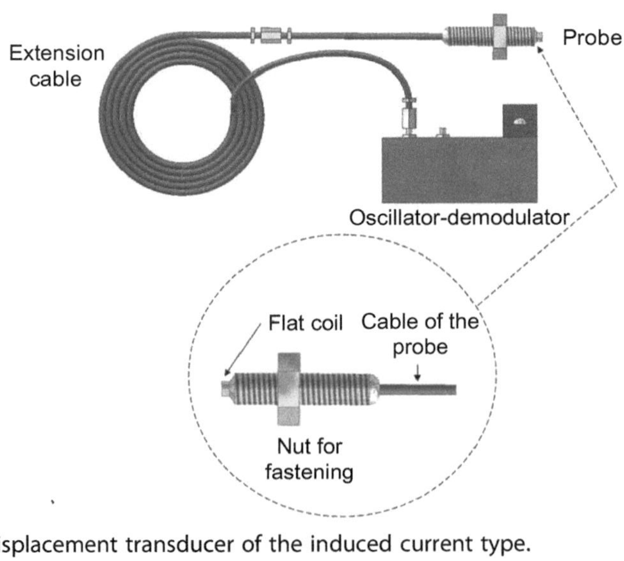


**Fig. 3.15** Displacement transducer of the induced current type.

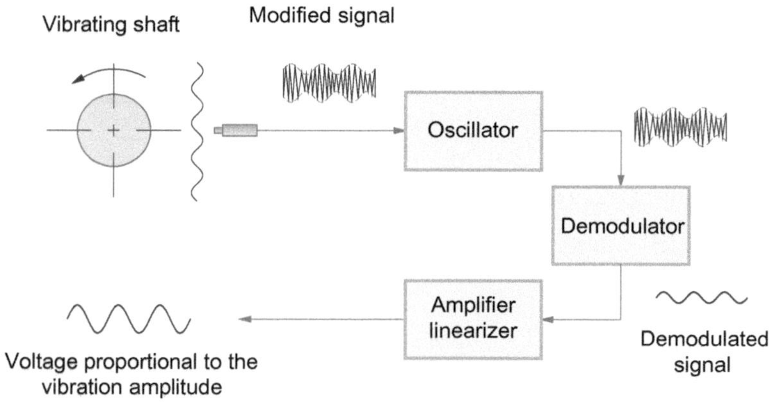


**Fig. 3.16** Processing of a displacement measurement.

that does not have the required length, then the transducer becomes uncalibrated.

Oscillator-demodulator: This devise supplies the required energy to keep the oscillations in the resonant circuit formed by the probe and the cable. The resonance frequency of this circuit is around 2.5 MHz. This signal is modulated in amplitude by current variations when increasing or decreasing the clearance. The demodulator allows separating the signal envelopes, as can be seen in Fig. 3.16. The resulting signal is sent to a linear amplifier, which allows obtaining a voltage signal proportional to the measured clearance.

Most oscillator-demodulators require a negative voltage supply of $-24$ or $-18$ V, in such a way that the higher the voltage supply, the greater the linear range of the transducer. The response for these types of transducers depends to a great extent on the type of material from which the rotor to be monitored is made. Therefore, every oscillator-demodulator must have its calibration curve.

The measurements of displacement carried out by this type of transducer in rotating equipment are normally expressed in mils, meaning thousandths of an inch (0.001 inch = 0.0254 mm).

The great advantage of these sensors over the others described is their application to low frequency measurements and their great temperature stability.

CHAPTER FOUR

# Causes and effects of vibration

## Monitoring machinery parameters in plant

One of the fundamental requirements to achieve optimal predictive maintenance is identifying the least expensive methods to monitor critical operational parameters of equipment, machinery, and systems to prevent possible future failures. In order to identify these parameters, it is necessary to know the overall functioning of the equipment, machinery, and the most common systems of an industrial plant. It is useful to group them into mechanical, electrical equipment, or related to heat transfer and/or hydraulics because each of these pieces of equipment has intrinsic operating characteristics in its design and operating conditions.

In general, the mechanical systems are composed of drive units, couplings, and final actuators. All of them have parts in motion such as shafts, reciprocating mechanisms, cams, gears, pulleys, and bands. For example, from drive units there are electric motors, gas turbines, steam turbines, and internal combustion engines. In coupling elements, there are couplings, gearboxes, bands, chains, and ball bearings. Final actuators are pumps, ventilators, conveyors, mills, and process equipment. In this way, attending to common characteristics in all the aforementioned equipment, the most-used method in monitoring their operation is the recording and analysis of their vibrations. The main sources of excitation of these mechanical systems come from the dynamic forces generated by the elements in motion of the electrical and magnetic fields, or from the ones induced by fluids.

The relation between the velocity and load define the operating conditions of a machine. Its operation can occur under constant velocity and constant load conditions, constant velocity and variable load, variable velocity and constant load, or variable velocity and load in such a way that, depending on its operation mode, the frequency and amplitude of the vibrations generated will be defined. Depending on these conditions, each machine component of the studied mechanical system will generate particular excitation forces that will characterize it.

*Mechanical Vibrations and Condition Monitoring*
https://doi.org/10.1016/B978-0-12-819796-7.00004-4

Due to the mechanical joints among machine elements, excitation forces are transmitted to other components. These transmitted forces causes vibration, so measuring this vibration in specific points gives information about the machine's set operating conditions. Amplifying these excitation forces, the vibration frequency spectrum is also amplified, therefore making it possible to observe signals related to machine component failure.

This chapter describes the most common vibration sources and analyzes how the spectra are amplified (increase in vibration amplitude) and expanded (more frequency components appear) in such a way that frequency signals on the spectrum related to machine component failure can be identified.

# Causes

## Imbalance

This occurs when the mass center of an equipment-rotating axis does not correspond to the axis geometric center. This situation generates a harmonic force acting perpendicularly to the rotating shaft. In this way, such force produces vibration due to the inherent elasticity of the mechanical systems.

The main causes of imbalance are related to manufacturing defects, wear of the ball bearings, misalignment, shaft bending, thermal deformations, and machinery foundation failures. The force generated for these causes can be represented as $F_D = mr\omega^2(\text{sen}\ \omega t - \phi)$, where $F_D$ is the excitation force, $m$ is the mass of the disc to be considered, $r$ is the distance between the gravity center and the rotating shaft, and $\omega$ is the angular rotor velocity and the phase angle.

So, the imbalance that a rotor presents depends on the relative position of its center of mass with respect to its supports. In the terminology related to predictive maintenance, the position of the center of mass is known as the heavy point, represented schematically in Fig. 4.1. Depending on the position of this point, there could be a difference in the position in which the maximum amplitude ($mr\omega^2$) or the maximum imbalance force ($F_D$) is detected.

A very important quantity for vibration analysis is the measurement of the phase angle ($\phi$) between two points of the shaft support. In case the angle is measured in the two supports of the shaft in the same direction, a stable value of this angle will be obtained if the rotating velocity ($\omega$) remains constant. Likewise, the amplitude of vibration will remain practically unchanged if measured in any radial direction to the rotating shaft. The difference in

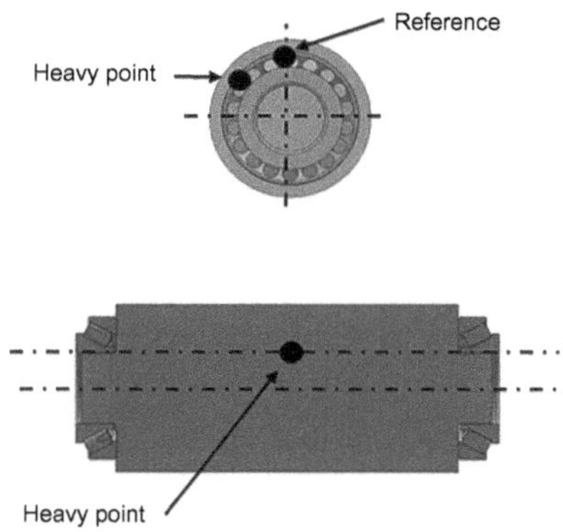


**Fig. 4.1** Schematic representation of the heavy point concept in a rotor.

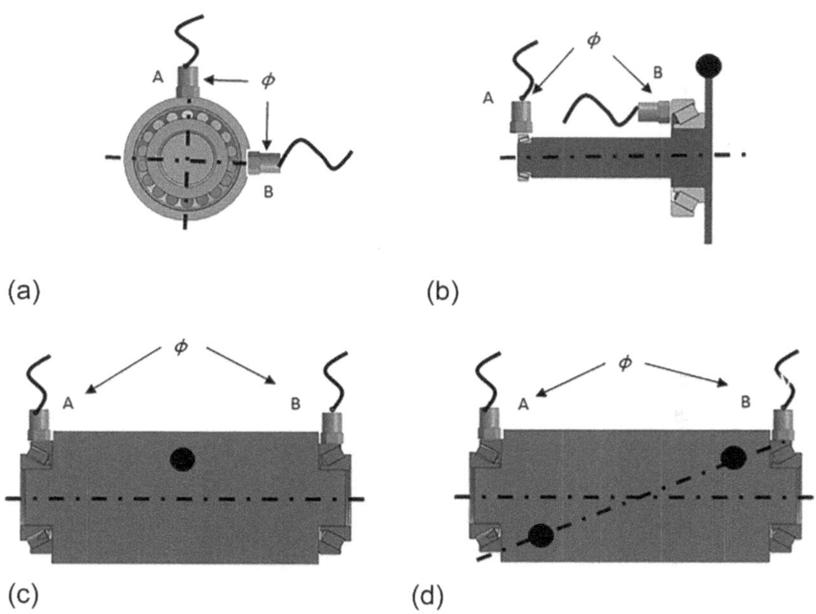


**Fig. 4.2** Location of transducers for the phase angle measurement in a rotating shaft. (a) Perpendicular on the same support. (b) Radial in support A, axial in support B. (c) Same phase angle in both supports. (d) Out of phase in both supports.

amplitude that could be present is due to the rotor support radial stiffness variation. Fig. 4.2 presents the most common location of transducers to register the phase angle in rotating equipment.

In Fig. 4.3, a typical frequency spectrum is shown for an imbalance case. In this spectrum, there is a peak corresponding to the shaft rotation

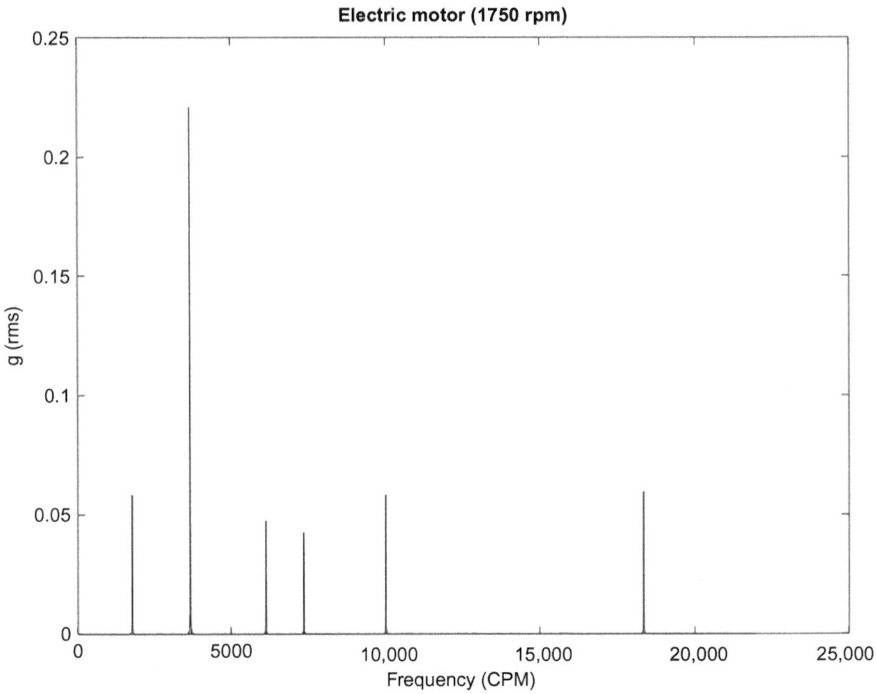


**Fig. 4.3** Frequency spectrum for a case of imbalance.

frequency (1$x$). In the case of a shaft with various discs, each with its heavy point, the shaft rotation will have a frequency spectrum that will present signal peaks when the rotation frequency coincides with the natural frequencies of each disc, as can be seen in Fig. 4.4.

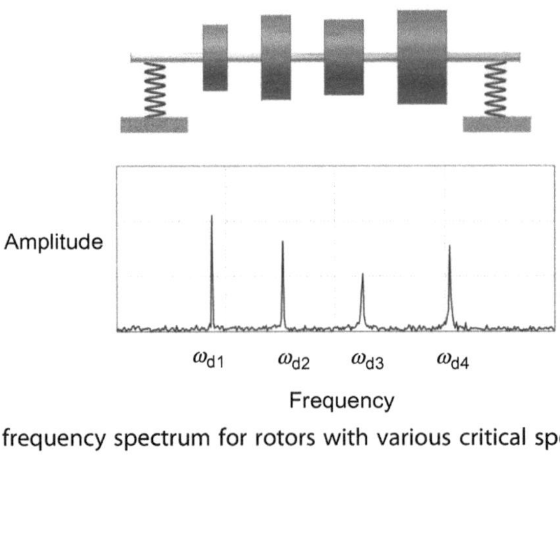


**Fig. 4.4** Typical frequency spectrum for rotors with various critical speeds.

In cases in which the imbalance is high, vibration also appears axially. In this case, in the frequency spectrum a peak will appear to a multiple or sub-multiple frequency value of the rotational velocity. A particular case of imbalance is presented when mounting the rotors in a cantilever, as can be seen in Fig. 4.2b. Unlike the rotors mounted between two supports, the ones mounted in only one support present axial vibrations that can be of greater amplitude than the radial vibrations.

Imbalance is especially serious when the rotation frequency coincides with one of the natural frequencies of the system, producing the resonance phenomenon. Likewise, when the vibration amplitude produces deformations beyond the shaft material fatigue level of the mechanical system components, premature failure occurs.

There are other vibration forms that show peaks in the spectrum to the shaft rotating frequency, and so their frequency spectrum is very similar to the one of the imbalance. For example, a shaft can be balanced and rotate eccentrically, a situation that produces a highly directional vibration, that is, if the amplitude is measured between both the horizontal and vertical directions, it will be found that the values are very different and the phase angle between these two measurements will be very close to 180 degrees.

Shaft bending can occur by thermal expansion or due to the shaft's weight, as is the case with rotors of the large turbo-generators such as that of the turbine shown in Fig. 4.5. The vibration that produces shaft bending mainly acts in the axial direction to the rotating shaft. In this case, in a frequency spectrum, one peak will be observed at the frequency of the second harmonic ($2x$) and another of the lower amplitude to the rotation frequency ($1x$). The phase angle measured between two consecutive supports will be around 180 degrees, as can be seen in Fig. 4.6.

In the shaft of big turbo-generators composed of several rotors, bending is avoided by keeping the rotor rotating at low speed. There are cases in

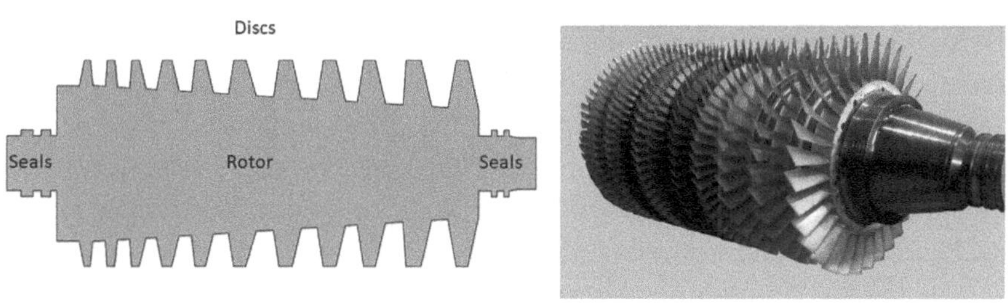


**Fig. 4.5** Gas turbine rotor.

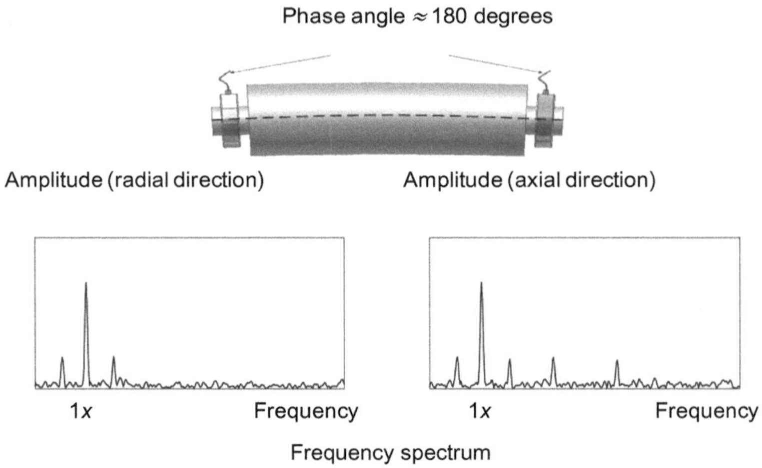


**Fig. 4.6** Phase angle and typical spectra in both the transversal and axial in a bent shaft.

which excessive bending is produced by the unbalance load, and it recovers once the shaft is in a static condition. This situation might appear in shafts operating at velocities close to their critical one. In electric motors, shaft bending might occur due to an electrical problem in the windings, so a rise in temperature in a localized area could deform the rotor shaft by thermal expansion.

## Misalignment

The lack of colinearity in the coupling of two shafts or between a shaft and a ball bearing is called misalignment. Three types of misalignment can be identified: the angular, the parallel, and the one caused by ball bearings, which are illustrated in Fig. 4.7. These types of misalignments produce both an axial and a transversal load that in time produces vibration of the shaft in these directions. The vibration frequency caused by these loads presents at the shaft rotation frequency ($1x$) and to multiples of it ($2x$, $3x$, etc.).

Angular misalignment is characterized for presenting a high vibration level in the axial direction of the coupling to the fundamental rotational frequency ($1x$) and to the second harmonic ($2x$). The amplitude of the second harmonic is approximately 30% higher than the one of the vibration to the fundamental frequency, and the phase angle between both sides of the mechanical coupling is around 180 degrees.

The parallel misalignment produces a higher vibration amplitude transversally to the rotating shaft. The frequency spectrum typical in this type of

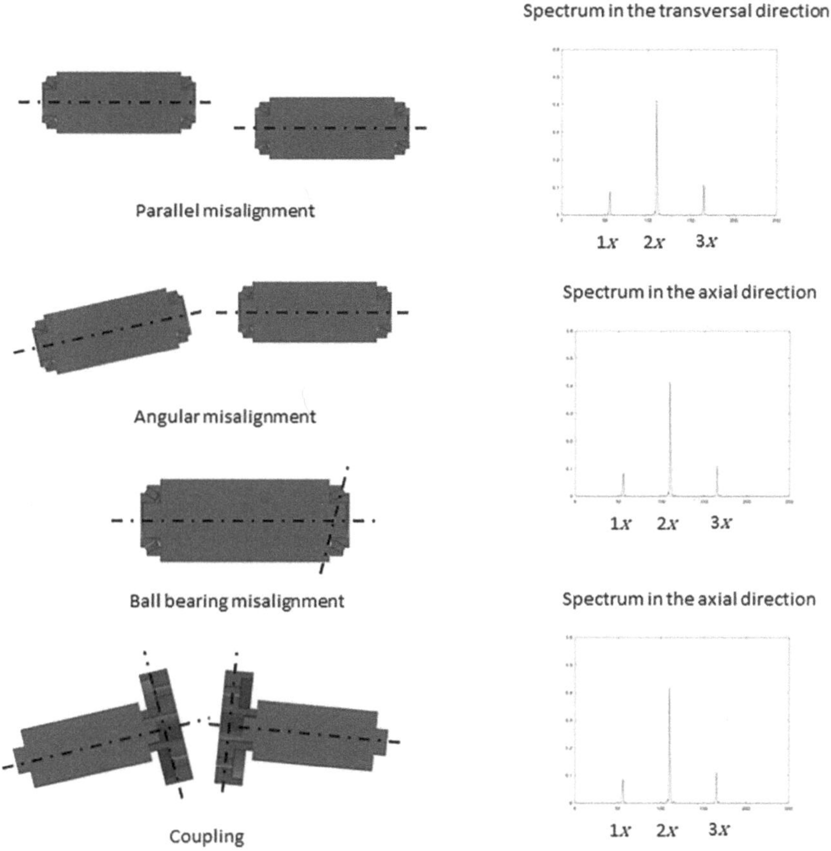


**Fig. 4.7** Schematic representation of the misalignment, their frequency spectra, and the phase angle.

misalignment presents peaks to the fundamental frequency ($1x$) and to the second harmonic ($2x$), with the signal amplitude of the second one being approximately 50% higher than the amplitude of the corresponding peak to the fundamental frequency. In this case, the axial vibration is of very low amplitude.

Ball-bearing misalignment appears due to improper bearing assembly and produces a vibration mainly in the axial direction to the rotating shaft. In this case, in the frequency spectrum, peaks to the fundamental frequency and to the second harmonic ($1x$ and $2x$) also appear predominantly. The value of the phase angle between diametrically opposed points will be around 180 degrees, as can be seen in Fig. 4.8.

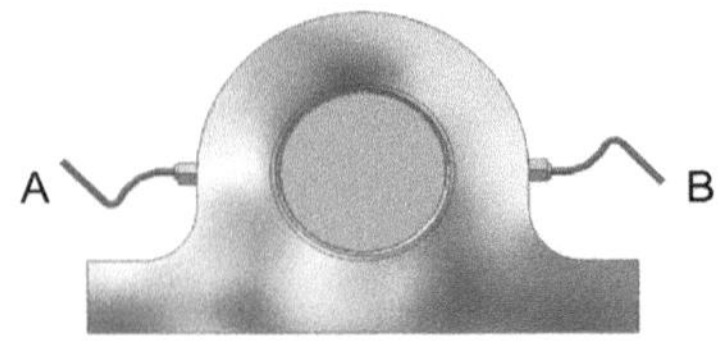


The vibration amplitude in A and B is similar
The phase angle will be around 180 degrees

**Fig. 4.8** Measurement of both amplitude and phase in a ball-bearing misalignment.

Ball-bearing misalignment is hard to distinguish from angular misalignment, so it is advised to uncouple the machine and verify the ball bearing independently. For all cases, if misalignment is too high, more harmonics ($3x$, $4x$, ...) of the fundamental frequency ($1x$) will appear in the frequency spectrum.

Most of the problems related with vibrations do not appear separately. It is tough to find signals that only contains information about a single problem, for example, only misaligned or only unbalanced. Moreover, the frequency spectra of two different problems might be very similar. This situation can be analyzed using vector diagrams. In these diagrams, both the amplitude and the phase angle of the acceleration vibration signal can be represented in such a way that knowing the mass of the component of the machinery under study, the excitation forces with their magnitudes and directions can be represented. From an experimental point of view, in predictive maintenance, it is standard practice to start with solving the simplest problem and keep on monitoring changes in the frequency spectrum.

## Gears

Vibrations in gear systems are inherent to their operation. When gear teeth come into contact, forces are present and thus vibrations appear due to the clearance, or backlash, between the contact teeth. The greater the clearance, the greater the vibration amplitude in the gear assembly. Every time two teeth come into contact, an impact is produced and thus vibration appears. The frequency of this impact is known as gear frequency (GF) and is equal to the number of teeth times the angular rotor velocity, that is, $\text{GF} = Nx$, where $N$ is the number of teeth and $x$ the rotation frequency of the shaft under study.

When the teeth are in good condition and the gear is concentric, the gear frequency is of low amplitude. If the teeth are worn, the envelope that generates the contact points between gears is distorted and the shape of the vibratory wave is no longer harmonic. Gear teeth profile deformation excites both the second and third harmonics of the gear frequency (2GF and 3GF), so monitoring these frequencies is a good advance indicator of teeth wear.

When the gear mesh is eccentric or the mesh problem is located in some teeth, the gear frequency gets modulated, as can be seen in Fig. 4.9A. Sidebands appear in the frequency spectrum, as shown in Fig. 4.9B. Spacing of the side bands is equal to the rotation frequency of the shaft that has the damaged gear. From the frequency spectrum, the shaft that produces more vibration can also be identified because the highest amplitude peak will be present at the frequency of rotation of the shaft with the damaged gear.

Measurement of gear vibration in the time domain is very useful to determine teeth condition. On the one hand, if the form of the measured signal is harmonic, this indicates that the gears are in good condition. On the other

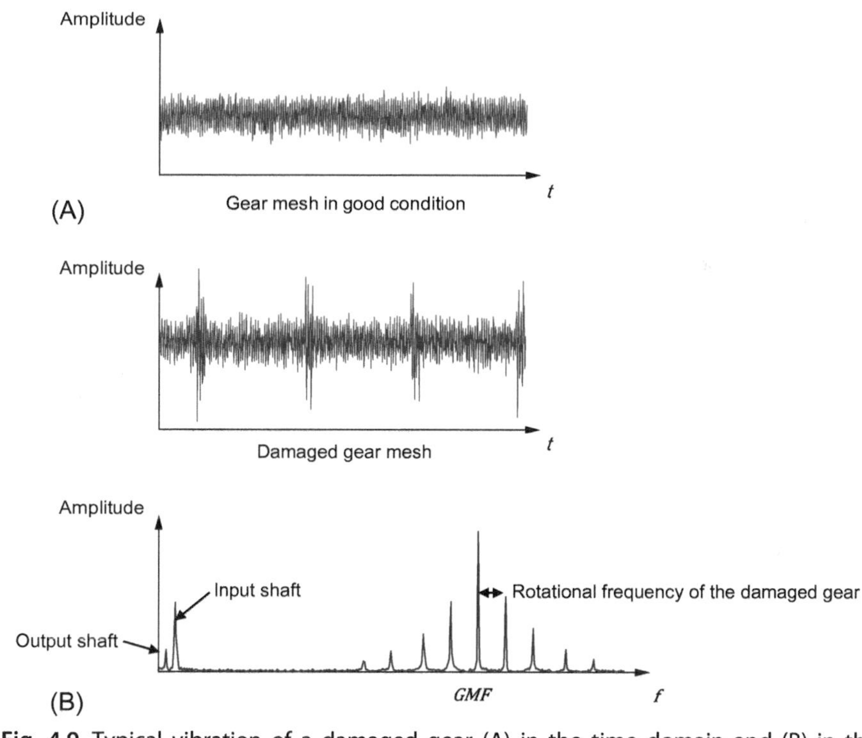


**Fig. 4.9** Typical vibration of a damaged gear (A) in the time domain and (B) in the frequency domain.

hand, if the signal is modulated, there is a problem in the gear mesh. A frequency spectrum regarded as normal must present peaks at the rotation frequencies of both shafts of the gear train and to the gear frequency (GF) of each of the gears in contact.

For a proper analysis of the signals observed in the frequency spectrum of a gear mesh, it is required to consider that the presence of peaks of great amplitude in the spectrum does not necessarily indicate damage in the gears because these signals are very sensitive to load variations. However, if sideband signal amplitudes are relatively high with respect to the gear frequency signal (GF), this suggests that the gear mesh is damaged.

The surface finish of the gear teeth is of great importance regarding the noise produced by the gear mesh. This noise is related to a high-frequency and low-amplitude vibration. Another important factor that influences the dynamic response of a gear train is the lack of concentricity in the teeth profile generation. This situation produces peaks of multiples of the fundamental mesh frequency in the frequency spectrum. These effects are more noticeable if any of the natural frequencies of the gear train are excited.

A vibratory firm typical of a gear mesh presents a peak to the gear mesh frequency with a series of sidebands and symmetric peaks separating each other to a distance equal to the rotor velocity, as indicated in Fig. 4.10.

From the predictive maintenance point of view, the analysis of low-frequency gear mesh vibrations is more useful than high-frequency analysis. This is because low frequencies are related to manufacturing errors such as teeth pitch error, teeth profile generating error, or teeth flexibility that basically generate noise.

As a summary of the concepts mentioned in this section, Fig. 4.11 shows the typical spectra of gear teeth wear, excessive clearance in the gear mesh, and gear misalignment.

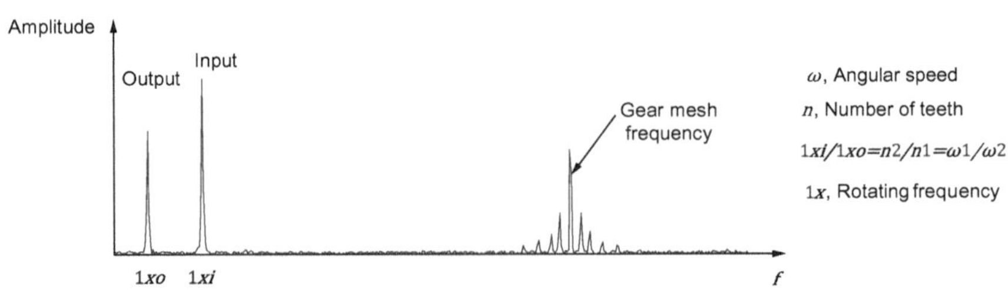


**Fig. 4.10** Main operating parameters of a gear mesh.

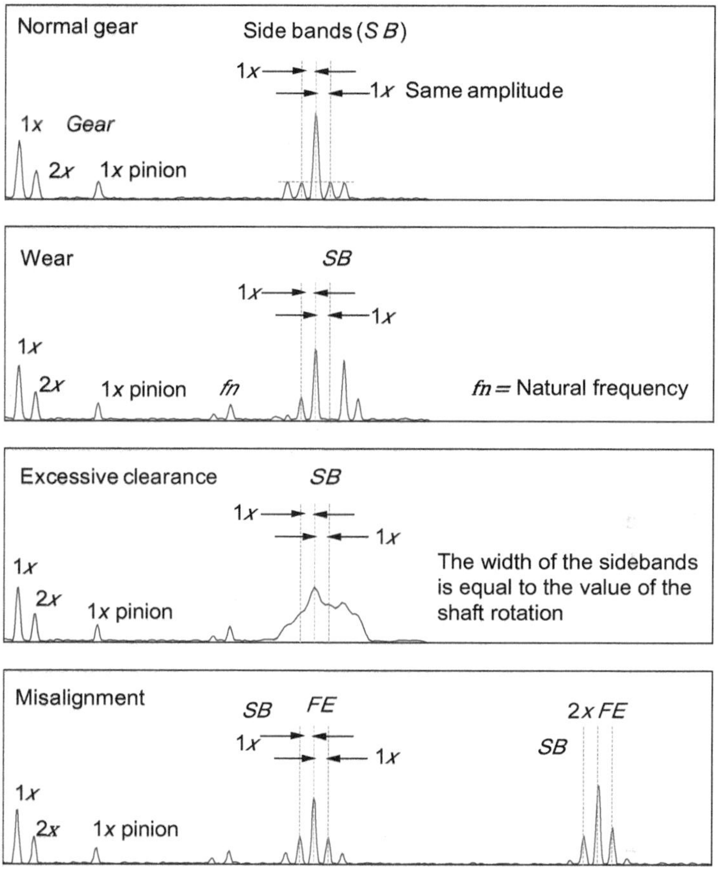


**Fig. 4.11** Typical spectra of faults in a gear.

## Journal bearings

Their function is to provide a contact point between a moving part and a static one through an oil film. Fig. 4.12 shows their main components, the shaft, the casing, and the oil film, and the acting forces over these elements from which the behavior of the oil film depends, and hence, the stability of the rotor in the journal bearing.

The operational principle of a journal bearing is based on an oil film that opposes a hydrodynamic force to the radial force produced by the shaft rotation. In this way, the shaft floats inside the oil film, thus shaft concentricity varies as a function of the oil pressure and temperature, oil properties, external load, rotation velocity, and operating conditions of the system.

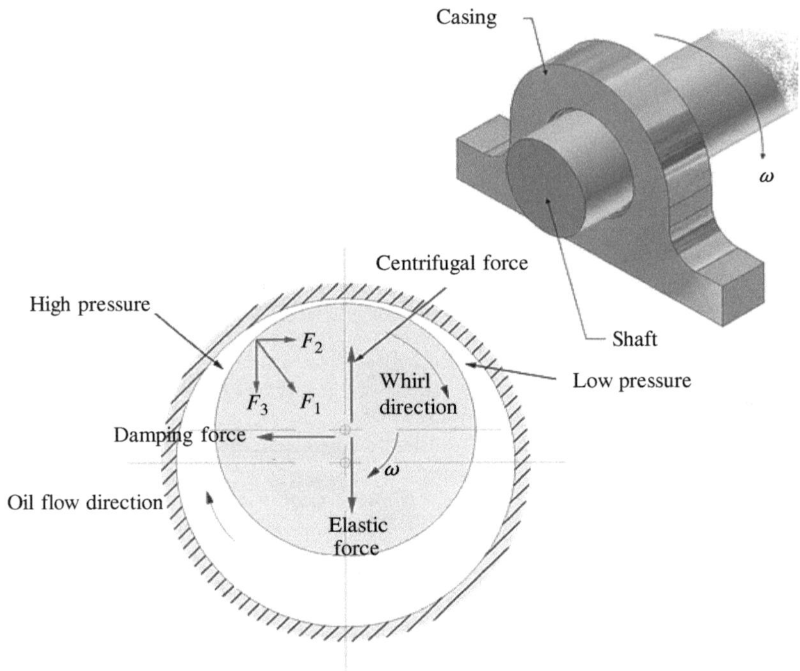


**Fig. 4.12** Acting forces in a journal bearing.

In steady conditions, the shaft keeps a position within a specified operational range according to the equipment design specifications. However, when a sudden change is present in the operating conditions, the position of the shaft is disturbed, producing vibrations at frequencies that are fractions of the fundamental one. Thus, the shaft goes into an unstable condition. In a journal bearing, two types of instabilities can be identified: whirl (oil whirl) and whip (oil whip). In the whirl instability, the shaft orbits around the housing of the journal bearing at approximately half the rotation velocity ($0.5x$). This motion arises due to changes in viscosity and oil pressure or by an external excitation force such as the one that could come from the vibration of nearby equipment. If the damping of the system is enough, the shaft will return to its normal motion. Otherwise, if the vibration increases, the oil film may break and produce metal-to-metal contact between the journal bearing housing interior side and the shaft. Whip instability occurs when the shaft is already on a whirl instability and the excitation frequency is close to a shaft natural frequency. This? type of instability is characterized by a shaft precession movement with respect to the journal bearing housing and could be catastrophic if not corrected in time.

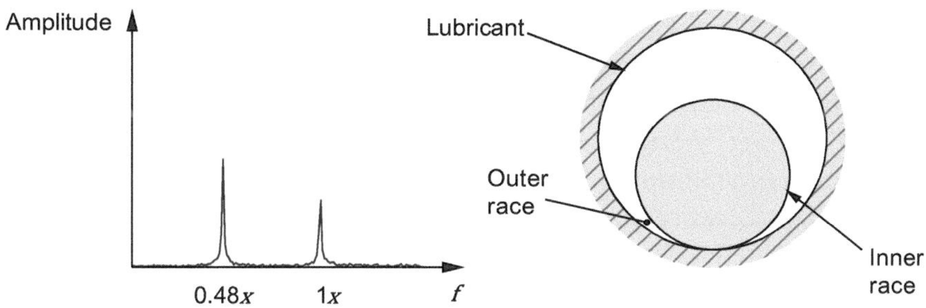


**Fig. 4.13** Typical spectrum of instability in journal bearings.

In the frequency spectrum, the journal bearing instability is present in the range of $0.4x$ to $0.48x$ ($x$, shaft rotation velocity), and it is considered very severe if its frequency reaches $0.5x$. Fig. 4.13 shows a typical spectrum of instability in journal bearings.

## Ball bearings and roller bearings

These mechanical elements, just like the journal bearings, allow rotating shafts to be supported on static structures, as can be seen in Fig. 4.14. In general, a bearing or ball bearing is composed of an inner ring with a very solid race and an outer ring with a race, whose surface is hardened, housing between the two rings a series of rollers, balls, or cones. The advantages that bearings present are their low friction coefficients, which allow the shaft to be accurately aligned, as well as their great resistance to momentary overloads, their simple lubrication, and the ability to support both axial and transversal loads.

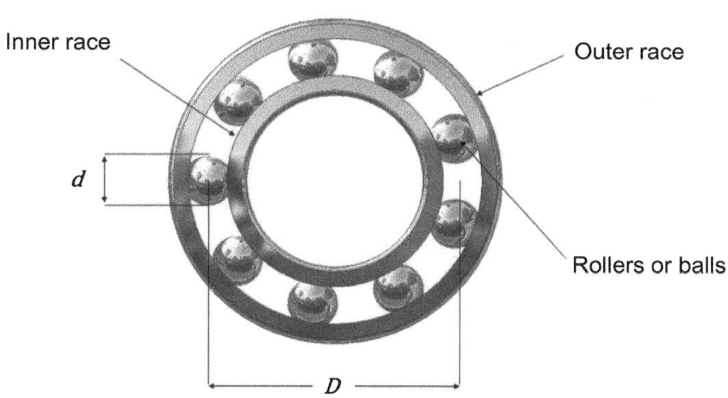


**Fig. 4.14** Typical components of a ball bearing.

The load capacity of a ball or roller bearing depends fundamentally on the resistance of the rotating elements to compression and wear. This capacity is associated with a predetermined period of useful lifespan, which is a function of its operating velocity. In this way, every bearing's moving element produces vibration associated with its rotation velocity. Thus, four basic frequencies are distinguished in their operation.

The rotation of the balls, which will be indicated as "BSF" (ball spin frequency); the one related to the race rotation (fundamental train frequency), which will be indicated as "FTF"; the one concerning the orbiting of the balls over the outer ring (ball pass frequency outer ring), called "BPFO"; and the one concerning the rotation of the ball over the inner ring (ball pass frequency inner ring), defined as BPFI. From the bearing geometry, the following expressions are obtained, allowing us to locate in the frequency spectrum the peaks related to each bearing element.

BSF frequency is associated with the rotation of the roller or ball over its own shaft and is calculated as:

$$\mathrm{BSF} = \left({}^{1}/_{2}\right)\left(1 - r^{2}\right)x \tag{4.1}$$

As the ball makes contact with two races in each revolution, a defect of the ball is 2BSF. The fundamental frequency of the bearing (FTF) is calculated as:

$$\mathrm{FTF} = \left({}^{1}/_{2}\right)(1 - r)x \tag{4.2}$$

A defect on the outer ring directly affects the ball spin frequency, so the ball pass frequency of the outer ring is given by

$$\mathrm{BPFO} = \left({}^{n}/_{2}\right)(1 - r)x \tag{4.3}$$

If the defect is located in the inner ring, it also affects the ball spin frequency, and the ball pass frequency of the inner ring is given by

$$\mathrm{BPFI} = \left({}^{n}/_{2}\right)(1 + r)x \tag{4.4}$$

In the above expressions:

$$r = \left({}^{d}/_{D}\right)\cos\alpha \tag{4.5}$$

$x$ = rotation frequency (Hz)
$D$ = bearing pitch diameter
$d$ = ball or roller diameter
$\alpha$ = angle of contact between the balls and races
$n$ = number of balls

In order to determine the ball spin frequency related to the races, when bearing dimensions are not available, it is common practice to use the following empirical relations:

$$\text{BPFO} = 0.4nx \text{ and } \text{BPFI} = 0.6nx$$

Bearings impose a lifespan on mechanical equipment, especially those that operate at high loads and velocities. Generally, the first vibration-related problems in a machine occur on the bearings, though a fault in this element could be the reaction to another problem in the machine. Generally speaking, bearings are the weakest elements in any rotating machinery. So, they affect the behavior of rotating shafts because the balls or rollers of which they are composed are elastic elements that can present minor differences in their construction and operation. This elasticity makes it difficult to define an absolute dynamic response. For this reason, their effects on a mechanical system are random, although they can be classified in four major groups: (I) defects related to the fundamental frequency of the mechanical system; (II) defects related to the ball or roller pass frequency in the inner ring; (III) defects related to the ball or roller spin frequency in the outer ring; and (IV) defects related to the ball or roller spin frequency. Each of these frequencies can be calculated and identified in the vibratory signature of each particular piece of equipment by using the relations previously presented.

Bearing vibration can be excited either by its own defects or by the reflection of induced vibrations on other machine elements. This vibration can be detected using narrow bands when carrying out spectral analysis of the problem.

The most common types of faults presented in bearings are the ball surface peeling, the superficial fatigue, and scuffing due to a slippage of the ring over the rotor or the housing. In particular, this fault presents a vibration spectrum similar to that shown by mechanical friction. Fig. 4.15 presents the stages by which a fault in bearings normally evolves. In the first phase of damage, the frequency spectrum reports a random vibration of high frequency in a range that goes from 1.2 kHz to 18 kHz. In the second phase of damage, a lower-frequency vibration gets excited (between 500 and 2000 Hz), but it is of greater amplitude than the high-frequency one. In the third phase, vibrations related to basic frequencies of bearing operation appear on the frequency spectrum, that is, FTF, BPFO, and BPFI can be identified. In the fourth phase, damage in the bearing is already very severe, presenting a considerable increase in the vibration amplitude of the fundamental frequency of rotation and the appearance of a great number of low-frequency components.

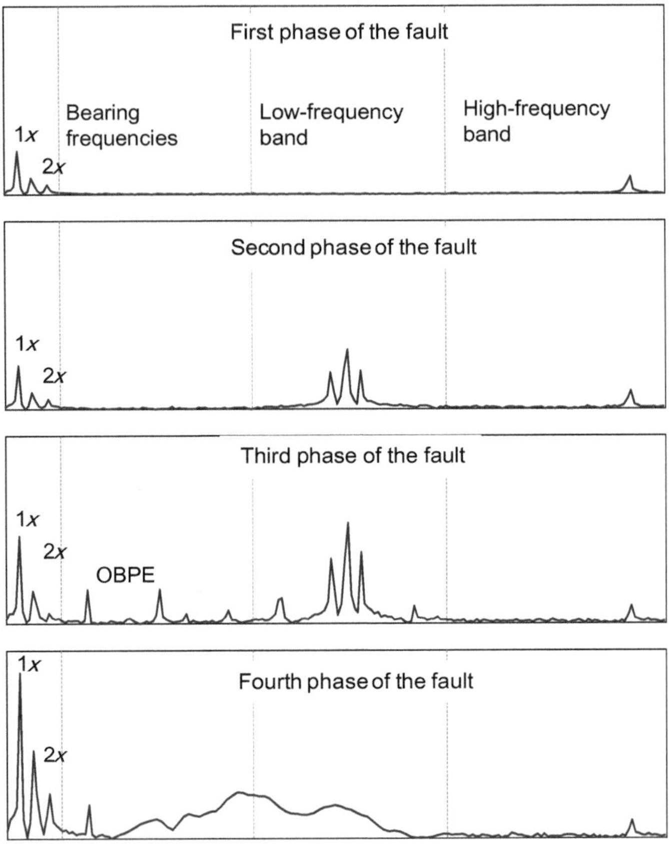


**Fig. 4.15** Evolution of the frequency spectrum with the increase in the severity of the fault in a bearing.

## Bands and belts

In band-based mechanical couplings, band wear, lack of tension, eccentricity, and misalignment of both the pulleys and the band itself are the main causes of vibration. All these causes of vibration in a pulley-band system are related to the band pass frequency (PF). PF as a function of the band pulley system geometry (shown in Fig. 4.16) is given by

$$\mathrm{PF} = \pi f_1 (d_1/L) \tag{4.6}$$

where

$$L \approx \pi(d_1/2) + \pi(d_2/2) + 2l \tag{4.7}$$

In these relations, if $f_1$ is given in Hz, PF will also be expressed in these units.

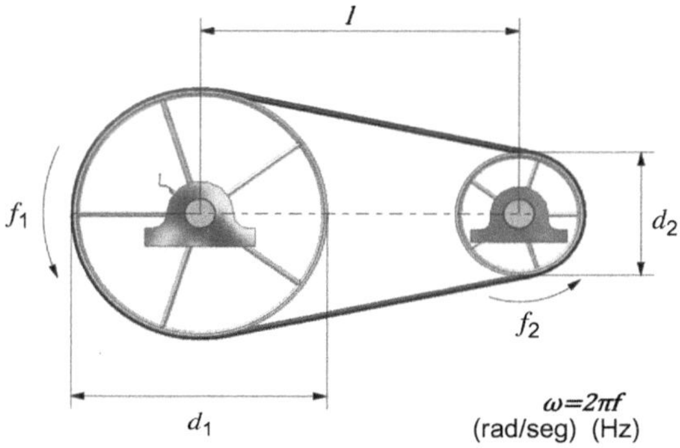


**Fig. 4.16** Geometry of a pulley-band system.

If the band is worn or loose, the frequency spectrum will show peaks at two and three times and of a lower amplitude at 4 PF. This vibration is in the transverse direction respect to the one of the shaft-pulley motion. In case a pulley is misaligned, peaks of greater amplitude will appear on the spectrum at the rotation frequencies of both the driven and driving pulleys. This motion will have its greater amplitude at the rotation shaft direction. If there is an eccentric pulley, the spectrum will show a peak at the rotation frequency of this pulley. The amplitude of this peak will increase noticeably when motion is measured in the band motion direction. That is, transversally to the direction of shaft rotation. In Fig. 4.17, a typical spectrum for a worn band is presented.

## Chains, cams, and mechanisms

Due to the great variety of mechanical elements and the new developments introduced in machine design, it is not possible to present an analysis of all of them. However, the methodology to analyze their motion is the same

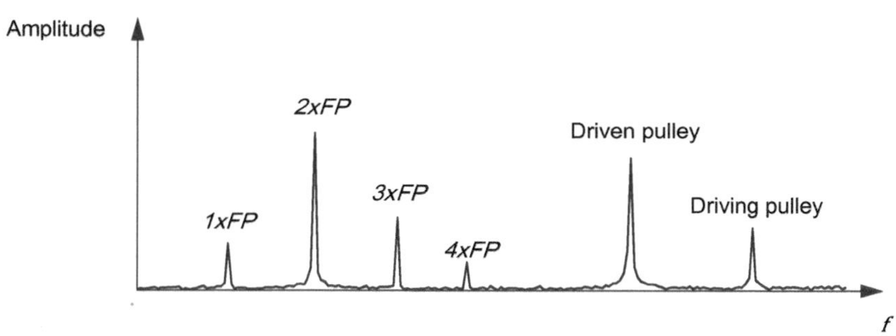


**Fig. 4.17** Typical spectrum of a worn band.

presented in the previous sections, that is, associating the peaks that appear in the frequency spectrum of the element, machine, or system under study, with its operating conditions as well as with the surrounding environment in which it is currently working. For example, chain sprocket-based transmissions produce vibration due to their geometry. In these systems, vibration arises due to clearance between the chain rollers with the sprocket teeth as well as for the deflection present on the sprocket teeth when engaged to the chain. In this way, vibration frequencies are associated with both the number of links that form the chain, the link dimensions, and the chain length between sprockets.

In the case of cams, conversion of a circular motion (cam) to an oscillatory one (follower) generates repetitive excitation forces that, combined with the elastic behavior of the mechanical system components, produce forced vibrations. In a frequency spectrum, both the cam rotational velocity and the follower oscillation frequency would be reflected. With these signals, it is possible to characterize the vibratory signature produced by the cam profile. Likewise, the articulated bar mechanisms transform motion in a mechanical system between an input motion that produces an output one, transmitting variable excitation forces through their supports. Measuring these motions and forces, it will always be possible to obtain the vibratory signature of the system and associate the operating conditions to the peaks observed in the frequency spectrum.

## Effects

The application of vibration analysis to predictive maintenance has a particular interest in determining the causes that produce excessive vibration. This section presents the effects that vibrations produce concerning faults, such as fatigue, loosening of pieces, friction between mechanical elements, noise, comfort, and safety.

Fatigue: In order to characterize the behavior of any working material subject to alternating loads (vibration), a graph is used that relates the level of stress applied to the number of load cycles. That is, a curve is formed, such as the one shown in Fig. 4.18, that represents how the material load capacity diminishes with the number of load cycles.

In a machine, the load cycles can occur completely under tension, compression, or a mixture of these loads, as can be seen in Fig. 4.19. The greater the vibration level to which a particular element of a machine is subjected, the shorter the time in which this element can operate under these conditions. That is, vibration amplitude directly affects the stress level.

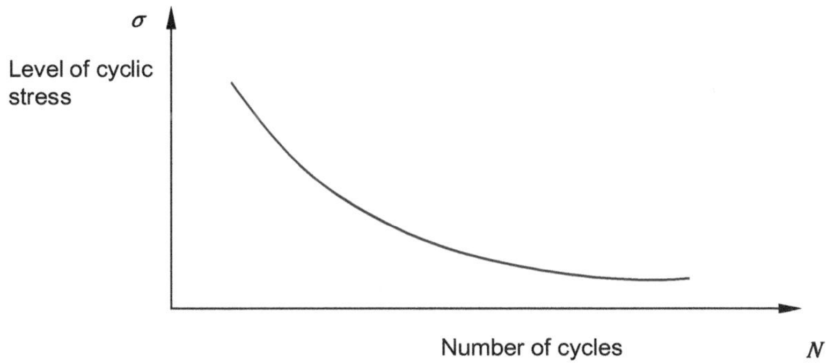


**Fig. 4.18** Typical curve of fatigue of the material of a mechanical element.

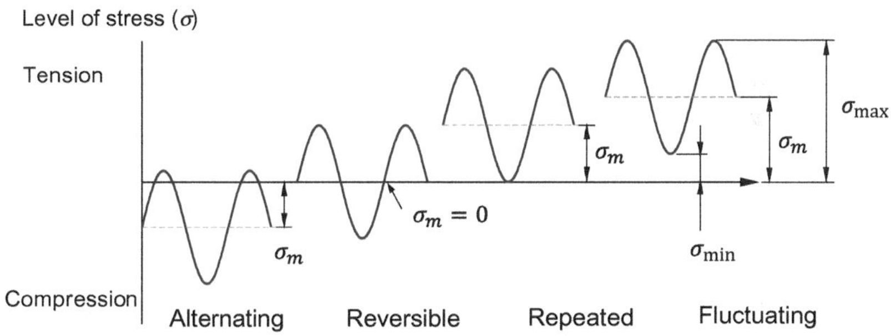


**Fig. 4.19** Variations in the application of the load-stress relation.

## Looseness

The loosening in a joint of machine components can be the result of a bad initial assembly or due to an excessive vibration level. However, regardless of the cause, once it starts, it increases, producing both misalignment and imbalance problems because its dynamic effect acts as a vibration amplifier. Fig. 4.20 shows a condition that reflects a loosening problem in a mechanical system.

In general, a loosening problem is nonlinear and random. Its causes may come from a bad foundation of the machine, the mechanical failure of some element, or bad assembly. Given this randomness as well as the diversity of excitation sources, a typical frequency spectrum of a loosening would present peaks at multiples and submultiples of the fundamental frequency of machine operation, as shown in Fig. 4.21.

## Friction

An excessive vibration amplitude causes friction among mobile machine elements to become a cause for catastrophic failure. Examples include a shaft that comes in contact with the Babbitt of its journal bearing, the rotor of

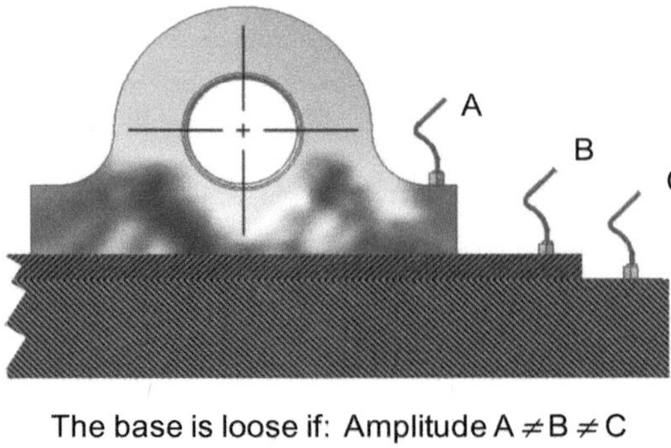


**Fig. 4.20** Loosening condition of the support of machinery.

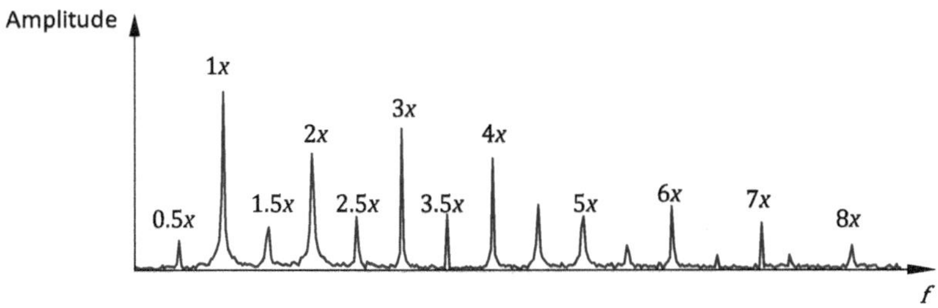


**Fig. 4.21** Typical frequency spectrum of a loosening problem.

an electric motor that makes contact with its stator, or the blades of a compressor or turbine making contact with the diffuser. Other examples of catastrophic friction problems caused by excessive vibration amplitudes are friction between a rotor with a seal or with the cover of coupling or a band or blades impacting their housing.

One of the predictive maintenance techniques that is feasible to diagnose troublesome friction consists of carrying out vibration measuring in a cascade. That is, registering the frequency spectrum at regular time intervals, thus being able to observe changes in the captured signal. For example, at the start of a machine, the cascade would show the variation of the vibration frequency depending on the shaft angular velocity. In the first start-up stages, generally the fundamental frequency dominates ($1x$); however, when there is friction, the vibration spectrum shows a peak at $0.5x$. This $0.5x$ peak in many occasions is of greater amplitude than the one corresponding to the fundamental frequency, as can be seen in Fig. 4.22.

In case of partial friction, this partial contact generates vibrations to subsynchronous frequencies (lower than the fundamental frequency), which in the frequency spectrum are identified as peaks of $0.25x$, $0.33x$, $0.5x$, etc. If

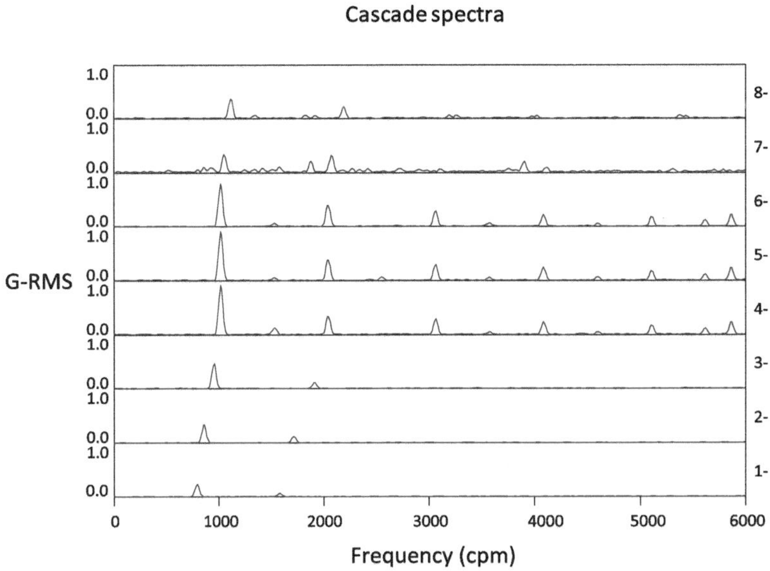


**Fig. 4.22** Spectra of sequential frequencies (cascade) when starting up a machine with a friction problem.

the partial friction occurs under a severe load, the subsynchronous frequency that dominates the spectrum is $0.5x$. If, in addition, the friction occurs along strong impacts, peaks will also appear at multiples of the subsynchronous frequency ($1.5x$, $2x$, $2.5x$, $3x$, $3.5x$, etc.)

In case of complete friction in a journal bearing the precession, the phenomenon can be presented in an inverse direction, thus the rotor will spin in one direction but it will orbit in the opposite one. This phenomenon is very unstable and can cause a catastrophic shaft failure. Another characteristic of this situation is that it starts at a resonance frequency and its signal on the spectrum remains independent of the rotation frequency.

## Noise

The noise produced as an effect of vibrations on machinery and within the context of predictive maintenance is of great importance because it affects the comfort and safety of the machine operators.

The most commonly used unit to measure noise level is the decibel (db), which indicates a relation between two quantities, and any of them can be a reference. In the case of noise measurements, the reference is the threshold of perception that a human being registers. Noise measurement is carried out registering an air disturbance, identified with the change of pressure that this disturbance exerts on an object, making it vibrate. The

**Table 4.1** Range of natural frequency of some components of the human body.

| System | Natural frequency (Hz) |
|---|---|
| Thorax-abdomen | 3 to 5 (standing person) |
| | 20 to 30 (sitting person) |
| Eye | 60 to 90 |
| Skull-jaw | 100 to 200 |

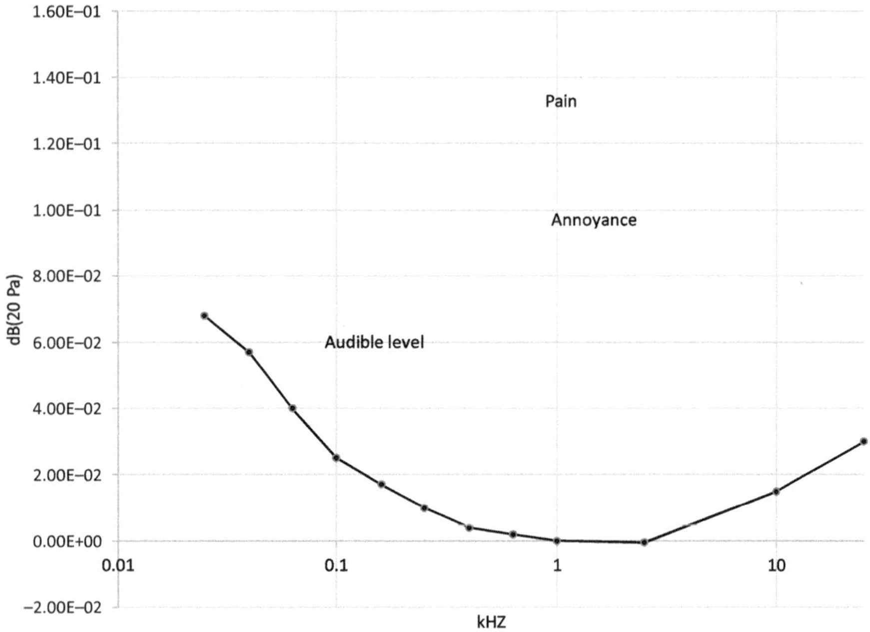


**Fig. 4.23** Human levels of noise perception.

threshold of perception registered by the eardrum of the human ear is around 20 micropascals (20 μPa), taking this quantity as the reference to measure the noise in dB.

Taking the dB reference for noise measurement, the relations between the generated noise level produced by machines can be established. The frequency and level of this noise can be defined and correlated with the noise perception of the machine operators. It is important to note that levels of noise perception vary from person to person because the natural frequencies of some components of the human body vary. A typical range of natural frequency values for human body parts can be seen in Table 4.1.

As a reference, Fig. 4.23 presents a graph that locates the noise level of different machines and the human perception of this noise level.

CHAPTER FIVE

# Alignment and balancing methods

## Introduction

The imbalance is the most common dynamic failure that affects rotating machines. It causes additional energy consumption, high levels of vibration, and noise. This failure is especially critical in equipment operating at high speeds because the centrifugal force generated can cause catastrophic damage. Reducing imbalance is of high industrial importance because it reduces operating costs; this chapter focuses on the techniques of detection and correction of imbalance, most used in the field.

The balancing of rotating equipment consists of trying to match up the center of mass of the rotor with the axis of rotation. There are different procedures for balancing a rotor. The most common methods are the vector method, the influence coefficient method, and the analysis method.

It is almost impossible to perfectly balance a rotor because the mechanical elements have manufacturing deviations, clearances, and backlash; even more, the shafts are elastic, and they bend during operation. It is practically impossible that the center of gravity of a rotor coincides with the rotation axis. Because it is impossible to eliminate the imbalance, there are balancing tolerances for each application. The standard ISO 1940-1:2003(E)[a] provides practical recommendations for balancing different types of rotors, and it defines a useful classification, as shown in Table 5.1, in which the maximum tangential speed ($V_m$) that the rotor can have is specified. The maximum eccentricity ($e$) that can support that rotor depends on the angular velocity ($\omega$). Table 5.1 determines the maximum velocity $V_M = \omega e$ ($e$ is distance from the rotating shaft to the center of the rotor's mass). Knowing the mass of the

[a] The new version of this standard is ISO 21940-11:2016.

*Mechanical Vibrations and Condition Monitoring*
https://doi.org/10.1016/B978-0-12-819796-7.00005-6

**Table 5.1** Classification of the balancing methods, according to the maximum tangential speed $V_m$.

| Quality levels of balancing[a] G-$V_m$ | Typical cases of application |
|---|---|
| G-4000 | Diesel engine crankshafts for underwater use of lower speed |
| G-1600 | Crankshafts of four-stroke internal combustion engines of large size |
| G-630 | Diesel engine crankshafts for underwater use |
| G-250 | Crankshafts of four-cylinder diesel engines with a piston displacement speed greater than 10 m/s |
| G-100 | Crankshafts of six-cylinder diesel engines with a piston displacement speed greater than 10 m/s, crankshafts of internal combustion engines, both gasoline and diesel, for automobiles, trucks, and locomotives |
| G-40 | Automobile wheels, rims, crankshafts of four-stroke gasoline or diesel internal combustion engines of six or more cylinders |
| G-16 | Rotors with motors, for example, propeller and cardan rotors<br>Agricultural machinery parts and grinding equipment. Engine components for automobiles, trucks, and locomotives |
| G-6.3 | Process machinery components, turbines for underwater use, centrifugal drums, ventilators, rotors of gas turbines, flywheels, pumping equipment boosters, machine tools, electric motors |
| G-2.5 | Gas and steam turbines, rotors of turbo-generators and superchargers, machine tools, electric motors with special specifications, turbo-pumps |
| G-1 | Recording systems, both tape and disc, disc player, lapping machines, high-speed electric motors |
| G-0.4 | Gyroscopes, discs and drill bits of high-precision lapping machines |

[a]The balance quality level represents the maximum tangential speed ($V_m$) that can reach the center of gravity of the rotor; it is given in mm/s.

rotor ($m$), the magnitude of the maximum centrifugal force ($F_m$) is also specified through the relation $F_m = m\omega^2 e$.

The permissible residual imbalance is determined from Table 5.1. It is defined as

$$U_{per} = \frac{(V_M)m}{\omega} \tag{5.1}$$

And the permissible residual specific imbalance is

$$e_{per} = \frac{U_{per}}{m} = \frac{(V_M)}{\omega} \tag{5.2}$$

The application of this method is illustrated with the following example. An electrical motor operates at a nominal speed of 1750 rpm (183.3 rad/s) and the mass of the rotor is 0.5 kg. According to Table 5.1, $V_M = 6.3$, therefore the permissible residual imbalance is:

$$U_{per} = \frac{(6.3)0.5}{183.3} = 0.01718\ (\text{kg mm})$$

And the permissible residual specific

$$e_{per} = \frac{(6.3)}{183.3} = 0.034\ (\text{kg mm/kg})$$

Or in grams

$$U_{per} = \frac{1000(6.3)0.5}{183.3} = 17.18\ (\text{g mm})$$

And the permissible residual specific

$$e_{per} = \frac{1000(6.3)}{183.3} = 34.36\ (\text{g mm/kg})$$

## Alignment

Before carrying out equipment balancing, it is necessary to ensure that it is correctly aligned. For this, it is required that the casing be well subjected to the supports and its coupling with the adjunct equipment be appropriate because two misaligned pieces of equipment produce vibrations in both machines as well as premature wear in the couplings, bearings, and seals.

One of the main causes of misalignment in a machine train is the incorrect assembly of the equipment; another is that it was aligned at room temperature, and when it began operating, the thermal expansion caused a misalignment of the shafts. Another important cause of this problem in high-torque machinery is the uneven wear of the couplings in such a way that the torque can misalign one of the machines. In general, these cases manifest when the equipment is operating and can only be detected by analyzing vibrations in a way that the higher the angular speed, the more critical the alignment would be.

The problem of thermal expansion must be accounted for with special care because the increase in the shaft direction is easily compensated, and it does not directly affect the alignment of the equipment; however, the transverse expansion significantly affects the position of the shaft in the space. Thermal expansion is a critical factor when aligning machine trains that utilize hot fluids, such as turbo-generators, in which the thermal expansion causes differential deformations in the casing and, as a result, the position of the shaft in the space varies in each ball bearing.

It is necessary to consider the arrangement of the mechanical train and to move only one device at a time. The alignment must be made in both horizontal and vertical planes. It is essential to avoid moving equipment that has connected pipes or those in which the foundation makes the operation more difficult.

The methodology for the alignment of any machine train comprises the following points: determination, in terms of alignment, of the optimal condition between the two pieces of equipment; the measurement of the relative position between the two machines to be aligned; calculation of the correction or movement that must be applied to the mobile machine to place it in the optimal location; and the movement of the machine to this position.

The relative position between the shafts of two machines is typically measured with dial micrometers (analogic), digital micrometers, and, more recently, laser-based equipment. When speed reducers are aligned, care should be taken with the flotation that the pinion suffers because the eccentricity of the gear as well as the axial movement of the system can indicate wrong measurements.

In some cases, the alignment can be done with the machines operating at normal temperatures. This procedure requires the measurement of the vibration levels while moving the equipment. The method could take longer, but the results are better than aligning when the machines are cold.

Equipment alignment is the less-expensive maintenance procedure and can reduce the vibration problems of many machine trains significantly.

## Balancing

The consideration of the balancing problem in the design and operation of all types of rotating or reciprocating machinery is fundamental to avoid harmful vibrations as well as to minimize the forces that the machine transmits to the supports and surrounding equipment. Balancing is the technique that allows eliminating the undesired inertia forces caused by machine elements (links, rotors, etc.) and that are directly related to operating speeds.

The manufacturing of machine elements is determined by considering the design tolerances. This situation defines the manufacturing cost, and it increases when the tolerances decrease; the tighter the tolerances, the higher the cost. Getting a proper relation between the cost and the level of vibration accepted makes necessary the application of balancing techniques for the correct operation of the machinery.

A fundamental concept in balancing analysis is the resonance that an unbalanced rotor can present. Fig. 5.1 shows a rotor supported on two bearings. In this case, it is supposed that the imbalance occurs on a disc placed at the distances $a$ and $b$, and where the centrifugal force is given by $F_m = m\omega^2 e$. In this way, the reactions in the supports are expressed as $F_1 = (F_m b)/(a+b)$ and $F_2 = (F_m a)/(a+b)$ and the movement of the center of gravity is driven by the following relation. The displacement of the center of mass was discussed in previous chapters and is defined as:

$$\overline{OA} = \frac{e(\omega/\omega_n)^2}{\left\{\left[1-(\omega/\omega_n)^2\right]^2 + (2\xi\omega/\omega_n)^2\right\}^{1/2}} \tag{5.3}$$

Assuming that the dampening is neglected, the expression would be:

$$\overline{OA} = \frac{e(\omega/\omega_n)^2}{1-(\omega/\omega_n)^2} \tag{5.4}$$

Here, $\overline{OA}$ is the amplitude of vibration as a function of the operating speed and natural frequency $(\omega/\omega_n)^2$.

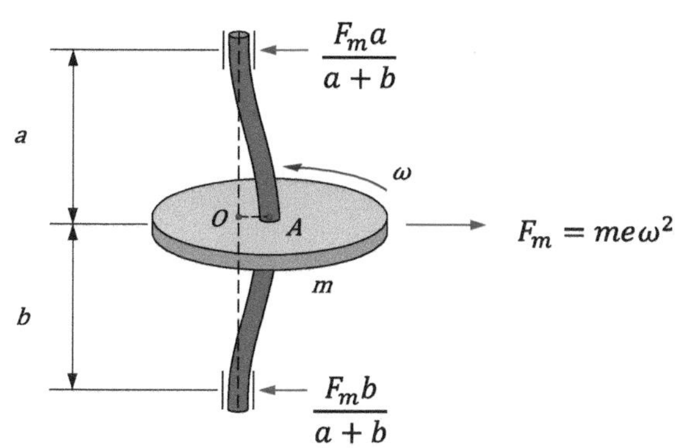


**Fig. 5.1** Schematic representation of the acting forces in an unbalanced rotor.

In the previous expression, it is observed that when the rotation speed $\omega$ is small in relation to the natural frequency of the system $\omega_n$ (critical angular speed), the amplitude of vibration is small. However, if $\omega$ increases, the amplitude of vibration increases until when $\omega = \omega_n$, the amplitude tends to be infinite. As the rotation speed is greater and moves away from the critical speed, the amplitude decreases, changing the phase that has the value of the eccentricity in a way that the center of gravity tends to agree with the rotating shaft defined by a line joining the centers of the two supports. To best illustrate the previous concepts, in Fig. 5.2 it was assumed that the geometric center of the disc ($A$) does not match with the center of gravity of the rotor ($G$), so $G$ tends to the position of $O$ when passing the critical speed.

In practice, the dampening ($\xi$) is not zero, so the increase of vibration amplitude to rotation frequencies ($\omega$) close to the critical speed occurs

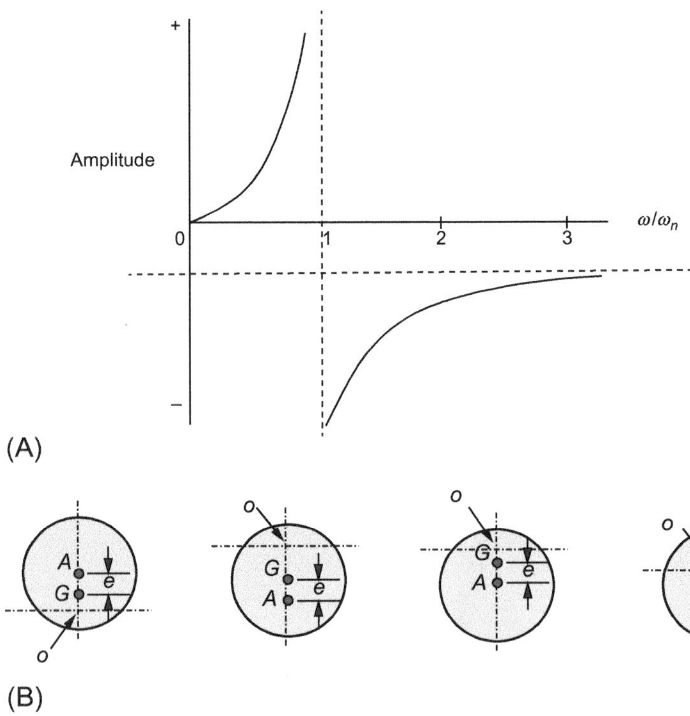


**Fig. 5.2** Representation of phase change when passing through a critical speed. (A) The amplitude as a function of the operating speed. (B) The relative position of the center of the disk $A$, the center of mass $G$, and the center of rotation $O$.

in a range of speeds. This is indicated by the expressions of Chapter One, which indicate that $\omega_d = \omega_n \sqrt{1 - \xi^2}$ where $\omega_d$ is the resonance frequency, considering the dampening of the system. The latter can be obtained through a logarithmic decrement test.

## Balancing by a vector analysis

Balancing a rotor is a method that adds or subtracts weight to move the center of mass the closest to the center of rotation. The first method for balancing a rigid rotor consists of vector analysis. This method sets the equilibrium equations and the sum of forces and moments in the rotor. When adding the weights of balancing, they could be equal to zero. That is, the rotor will be in balance.

The balancing by vector analysis is generally classified as static or dynamic. Static balancing, also known as single plane balancing, consists of balancing the rotor whose gravity center does not coincide with the axis of rotation. The rotor is represented as a simple system of particles rotating about a fixed point (representing the axis of rotation) and having a constant distance between the particle and the origin (fixed point). Fig. 5.3 illustrates this case. This rotor will be statically balanced if the sum of the momentums with respect to the origin is equal to zero, as can be seen in the same figure. However, it will be dynamically imbalanced because the sum of momentums with respect to the axes $Y$ and $Z$ is different to zero.

The static balancing only applies in the case of short rotors, that is, those in which the relation of their diameter ($D$) and length ($L$) is less than 0.5 ($D/L \leq 0.5$).

In the case of dynamic balancing, it is required to eliminate the inertia forces so that the acceleration of the gravity center of the rotor is equal to zero. The inertia forces depend on the angular speed of the rotor. For this, the balancing is carried out through a determined rate without ensuring that the rotor continues to be balanced to a different angular speed.

To carry out dynamic balancing, it is necessary to have two planes balancing because the momentums around the perpendicular axes to the rotation shaft of the rotor must be minimized, as can be seen in Fig. 5.4. If there were a single plane balancing, it would not be possible to minimize the momentum of rotation in the two supports of the rotor. Fig. 5.5 presents the vector representation of the sum of forces and momentums acting in the

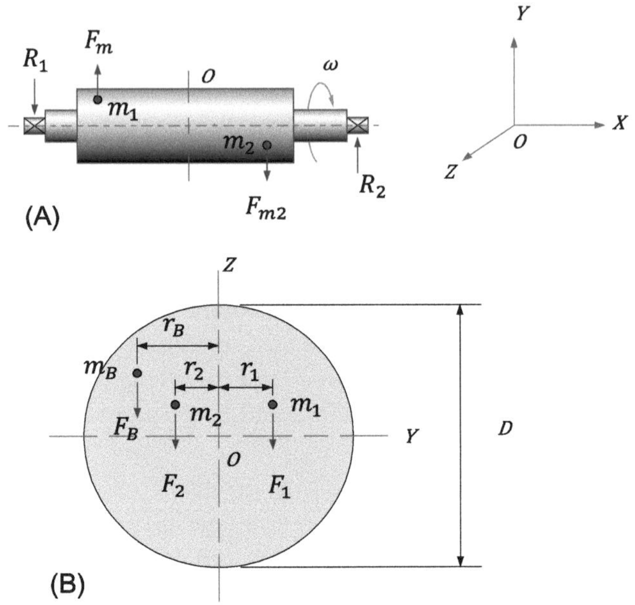


**Fig. 5.3** (A) Statically balanced rotor, but dynamically unbalanced, (B) for dynamically balancing the sum of momentums with respect to $O$ must be equal to $O$.

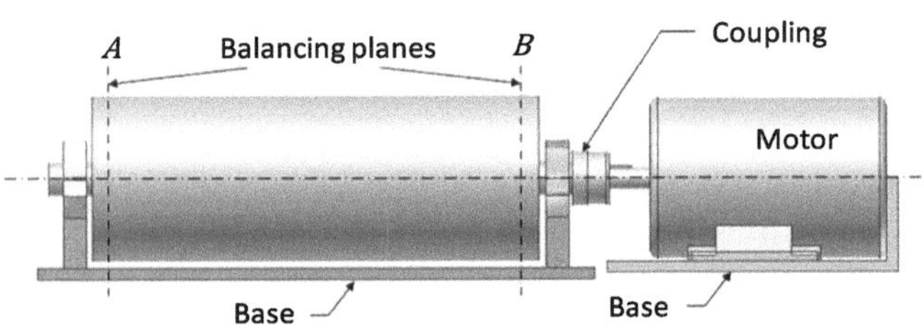


**Fig. 5.4** Typical location of the planes in a rotor for dynamic balancing.

rotor in a way that the following expressions can arise: $F_1 = m_1 e_1 \omega^2$, $F_2 = m_2 e_2 \omega^2, \ldots, F_n = m_n e_n \omega^2$, where $F_i$ con $i = 1, 2, \ldots, n$ are the inertia forces acting in the direction of the line that joins the center of the mass with the rotating shaft, being separated by a distance $e_1$; $m_i$ are the equivalent masses of the different discs of the rotor; and $\omega$ is the angular speed to which the balancing is carried out. For the balancing planes $A$ y $B$, it would be $F_A = m_A e_A \omega^2$ and $F_B = m_B e_B \omega^2$.

The balancing is carried out to a determined angular speed $\omega$, which is constant, and the inertia forces are proportional to the product $m_i e_i$. Breaking

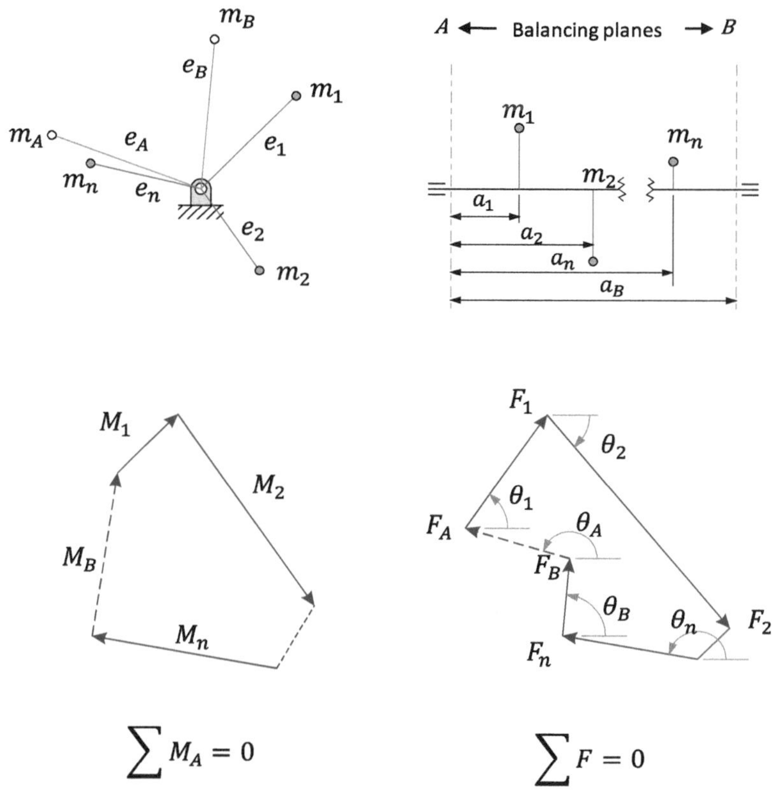


**Fig. 5.5** Vector representation of the sum of forces and momentums for the balancing of a two-plane rotor. Influence coefficients.

down the inertia forces in two orthogonal directions $X$ and $Y$, the sum of forces in the direction $X$ would be given by:

$$\sum F_X = m_1 e_1 \cos\theta_1 + m_2 e_2 \cos\theta_2 + \ldots + m_n e_n \cos\theta_n + m_A e_A \cos\theta_A + m_B e_B \cos\theta_B \quad (5.5)$$

and in the $Y$ direction it would be

$$\sum F_Y = m_1 e_1 \mathrm{sen}\theta_1 + m_2 e_2 \mathrm{sen}\theta_2 + \ldots + m_n e_n \mathrm{sen}\theta_n + m_A e_A \mathrm{sen}\theta_A + m_B e_B \mathrm{sen}\theta_B \quad (5.6)$$

The sum of momentums with respect to the balancing plane $A$ would be:

$$\sum M_{AX} = m_1 e_1 a_1 \cos\theta_1 + m_2 e_2 a_2 \cos\theta_2 + \ldots + m_n e_n a_n \cos\theta_n + m_B e_B a_B \cos\theta_B \quad (5.7)$$

Y, finally:

$$\sum M_{AY} = m_1 e_1 a_1 \text{sen}\theta_1 + m_2 e_2 a_2 \text{sen}\theta_2 + \ldots + m_n e_n a_n \text{sen}\theta_n + m_B e_B a_B \text{sen}\theta_B \quad (5.8)$$

From the expressions of the sum of momentums, the tangent of the angle $\theta_B$ can be obtained because all other parameters are known. Likewise, the value of the angle $\theta_A$ is obtained, if the sum of momentums is presented with respect to the balancing plane *B*. In this way, the angular position of the balancing weights in the planes A and B are determined. With this data, and using the previous four expressions, the values of $m_A$, $m_B$, $e_A$ can be obtained and $e_B$, suggesting this balancing problem is solved.

## Balancing in the field

The balancing of rotating machinery can be done in two ways: in the field or in a balancing jig. Balancing in the field is employed more in the industry because it can be carried out without disassembling the rotor, allowing the balancing of a machine in less time and at a lower cost compared to that of a balancing pit. Despite the aforementioned, it is necessary to consider that the accuracy of balancing in the field is lower than that in a balancing jig, although it is sufficient to get the specified operating conditions for most machines.

When applying the balancing method by a vector analysis, it was assumed that the imbalance masses ($m_i$) were known, as was their relative position ($e_i$) with respect to a reference line, but when a rotor is balanced in the field, normally, these data are not provided. That is why the so-called heavy points of a rotor must be determined experimentally. This is carried out by measuring the vibration amplitude and the phase angle to the operating speed of the machine with respect to an arbitrary reference line. The phase angle is measured with a tachometer trigger to the vibration measuring device or with a stroboscopic light.

If the balancing is made without measuring the phase angle, then the influence coefficient method can be used. That technique consists of adding test weights in the different angular positions in the rotor. Then, the influence that each of these weights has on the rotor is measured. The dynamic response of the rotor is characterized, and the location of the heavy point is localized. By subtracting the amount of weight, the rotor will be balanced. The application of the balancing method requires several test runs. The test run denotes the operation of bringing the rotor to its nominal operating speed (or at a speed that is needed for balancing) and carrying out both amplitude and phase measurements.

A simplified application of the influence coefficient method in a single plane is the so-called four tests run method. This method is very useful in the predictive maintenance context because it only needs a spectrum analyzer, avoiding the direct measurement of the phase angle. The method is applicable only when the operating conditions of the equipment to be balanced indicate that the dominant vibration occurs at $1x$, that is, to the rotation frequency. The analysis procedure consists of positioning the test weight at 0, 120, and 240 degrees with respect to an arbitrary reference. The test weight must be located at the same radius $r$. In the first test run, carried out without a test weight, the amplitude of vibration is registered, which is graphed, as can be seen in Fig. 5.6, as a circle that has a radius equivalent to the amplitude of vibration detected. For the second test run, the test weight is placed to 0 degree and the amplitude of vibration is measured again, drawing up a second circle on the graph whose center is the intersection of the circle that represents the original vibration with the reference shaft, as can be seen in Fig. 5.6. In the third test run, the test weight is placed at 120 degrees from the reference and the amplitude measurement is repeated. This latter is graphed as a circle with a center in the corresponding mark to 120 degrees in the corresponding circle to the original vibration. In the fourth test run, the test weight is placed at 240 degrees and the amplitude of vibration is obtained, graphing another circle, as can be seen in Fig. 5.6. This procedure is equivalent to solving a system of equations that results from obtaining the balancing weight and its phase angle in the intersection of the three circles corresponding to the vibration amplitude of the rotor, considering the different positions of the test weight.

## Modal analysis

The systems with various degrees of freedom are interconnected and the movement of each of them is dependent on the others. In the vibratory systems, each degree of freedom is associated with a natural frequency and, at the same time, each natural frequency produces a characteristic movement known as vibration mode. The ISO 11342 is the reference for balancing flexible rotors.

A system with several degrees of freedom, or a continuous system, will enter in resonance if the rotating speed coincides with one of the natural frequencies.

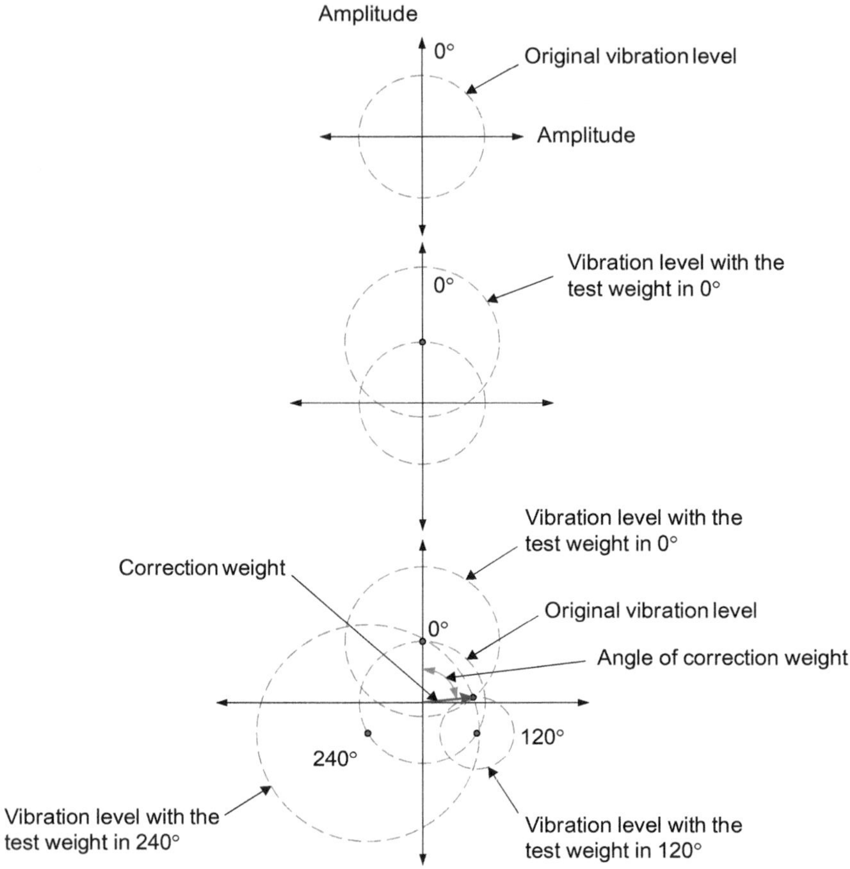


**Fig. 5.6** Balancing procedure of the four run tests.

If the system gets vibrated at different frequencies from those of the natural frequencies, its way of vibrating will correspond to a combination of all the system modes, with the vibration mode whose natural frequency is closer to the excitation frequency prevailing. This characteristic is the basis to carry out modal analysis. In this way, the vibration of a rotating system can be analyzed due to its critical speeds, that is, in their resonances, where the influence of the other vibration modes is practically null. This allows knowing the vibration mode by mode in such a way that with a good approximation, this analysis is equivalent to studying independent systems of a single degree of freedom.

Assuming that the rotating system in analysis is linear, that is, that the sum of the causes and effects of the vibration analyzed separately produce the same behavior of the rotor than when all these causes and effects act simultaneously, a rotor can be balanced by balancing each of its vibration modes separately. In the same way, the excitation forces of a system with a certain number of degrees of freedom can break down in an equal number of functions or excitation forces in such a way that each function or excitation force acts only on one mode without affecting the others and that the vector sum of all considered forces is equal to the original excitation function. In the case of balancing the rotors, these forces are the masses of imbalance to be compensated. Fig. 5.7 presents a rotor that can be analyzed as a system with three degrees of freedom represented as concentrated masses and indicating the shape that the vibration modes would take.

It should be noted that in the case of an imbalanced rotor, the vibration modes do not necessarily occur in the same plane; instead the modes will occur in different planes depending on the position of the imbalance in each disc. This analysis is more complicated because each disk has a vector component, and the vector sums have to be made at each mode considering both the overlapping and the location of the measurement planes of each mode.

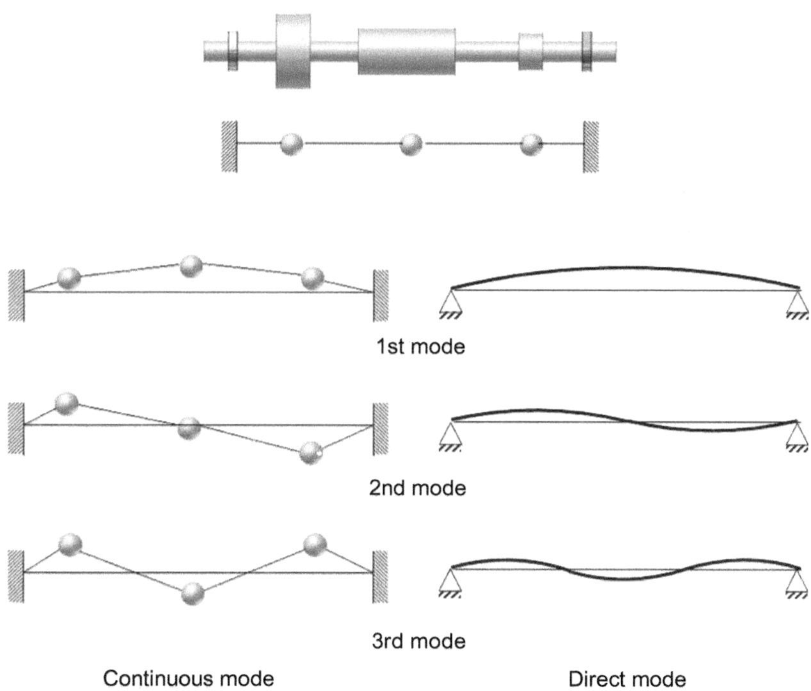


**Fig. 5.7** Vibration modes of a rotor with three degrees of freedom.

The use of polar diagrams to analyze these situations is very common where vector sums can be visualized.

Fig. 5.8 presents the rotor of a large turbo-generator, constituted by a series of elements of different diameter, mass, and stiffness, and where the oil film of the ball bearings acts as a spring (Fig. 5.8A). In this case,

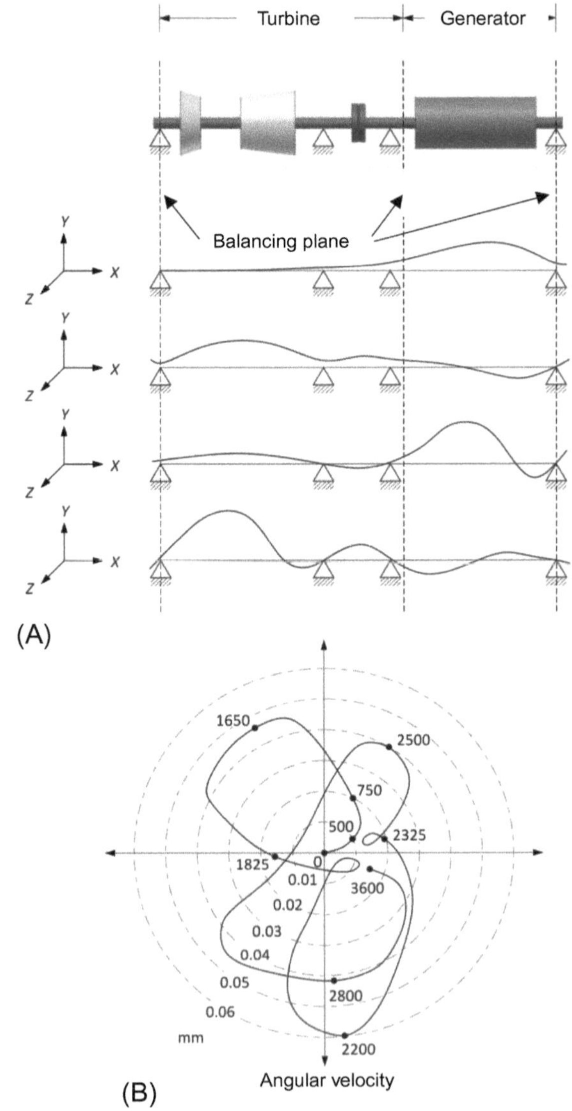


**Fig. 5.8** (A) Rotor of a turbo-generator, (B) deformations depending on the vibration modes, (C) polar diagram.

deformations of the rotor that do not occur in only one plane and that change with the different vibration modes are presented (Fig. 5.8B), The polar diagram of Fig. 5.8C, typical of a measuring point in a turbo-generator, indicates the evolution of the amplitude with the rotation speed in such a way that the critical speeds of the rotor can be identified. By associating the position of the measuring point with the amplitude and phase of vibration recorded, the vibration modes can be identified in approximate terms. With this information, it will proceed to minimize the amplitude of vibration of each mode in the balancing planes.

CHAPTER SIX

# Practical cases

## Introduction

In this chapter, the monitoring of equipment operating in industrial plants is presented. These examples illustrate the application of the concepts presented before and guide the reader to follow practical cases of machinery diagnosis. The diagnoses are based on the study of their vibration signals and spectra. This chapter complements the concepts described in the previous sections as well as presents their direct application in practical problems.

These cases cover diagnoses carried out in steam turbines, speed reducers, centrifugal pumps, ventilators, and machine tools. Every case describes the diagnose procedure, the interpretation of the vibration spectra, and the analysis of the root of failure.

## Steam turbine

The first case is the analysis of a turbine that shows high vibration levels during operation. It is necessary to have a real-time spectrum analyzer for the inspection and supervision of turbo-machinery; it is also necessary to have suitable accelerometers, pressure testers, and strain gauges. In this way, periodical monitoring allows recording enough data that input information for determining the tendencies of the different components to be identified as well as to predict possible failures. In this first case, an increment in the vibration levels of a steam turbine was found. As a consequence, a detailed measurement program was launched, and the vibration levels were measured at several points of the turbine by using the accelerometer. The data were analyzed with the Fourier transform. The problems the measurements detected were: High vibration in the shaft and the gears of the governor transmission as well as fractures on the blades of the first and third stages. The steam turbine had three stages that generate 4300 HP, and it moved a centrifugal air compressor.

*Mechanical Vibrations and Condition Monitoring*
https://doi.org/10.1016/B978-0-12-819796-7.00006-8

When the horizontal response was measured at the turbine inlet support, the first critical speed was presented at 4600 rpm (76.67 Hz) and not at 4800 rpm (80 Hz), as indicated in the design, and the pass frequency in the turbine nozzle was 3800 Hz. This is approximately 50 times the rotation speed.

The vibration signal was recorded at nine points along the machine train. Table 6.1 includes the frequency, direction, and corresponding modal shape.

According to Campbell's diagram, shown in Fig. 6.1, this frequency could excite the second, third, and fourth natural frequencies of the blades

**Table 6.1** Identification of the measuring points.

| Point | Frequency (Hz) | Direction | Mode |
|---|---|---|---|
| P1 | 4.0 | Radial | 3rd |
| P2 | 8.3 | Axial | 3rd |
| P3 | 13.3 | Radial | 1st |
| P4 | 18.0 | Radial | 3rd |
| P5 | 20.0 | Axial | 3rd |
| P6 | 25.0 | Axial | 1st |
| P7 | 33.3 | Axial | 3rd |
| P8 | 50.0 | Axial | 1st |
| P9 | 63.3 | Radial | 3rd |

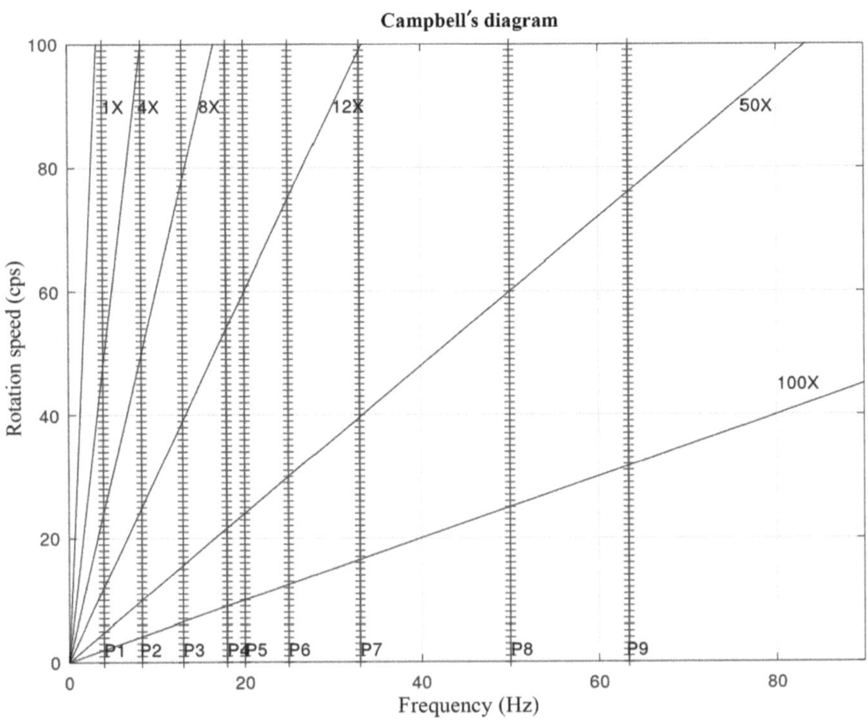


**Fig. 6.1** Campbell's diagram, typical in a steam turbine.

that are located at the first and third stages of the turbine; this situation was avoidable if the turbine were rotating at its operational speed of 80 Hz. This situation is evident because the intersection between the 50x line and the operating speed could be avoided (x is the nominal speed, for this case 50x= 4000 Hz). The turbine would be operating in a safer condition.

Campbell's diagram of a turbine provides the relation between the natural frequencies of the blades with its own operating speed. The intersections between the lines, corresponding to both the operating speed and the natural frequency, represent the resonance frequencies.

Normally, a turbine fails due to design errors, bad installation, or inadequate operation. The most vulnerable components are the ball bearings and the blades. The latter generally get fractured near the root and the type of fracture is, as in this case, due to fatigue caused by vibrations at close frequencies to the natural frequency of the blade. When studying the steam leaks in the periphery of the turbine nozzles, detected with an ultrasound sensor, pressure variations causing partial admission effects were identified and, with this, the variation of the operating speed.

## Bearing in a speed reducer

An increase in both the temperature level and the noise produced in the speed reducer of a mill was reported. The machine train arrangement is shown in Fig. 6.2. The figure shows the location of the gearbox and the measuring points. Table 6.1 describes the operating conditions and the design parameters of the main components.

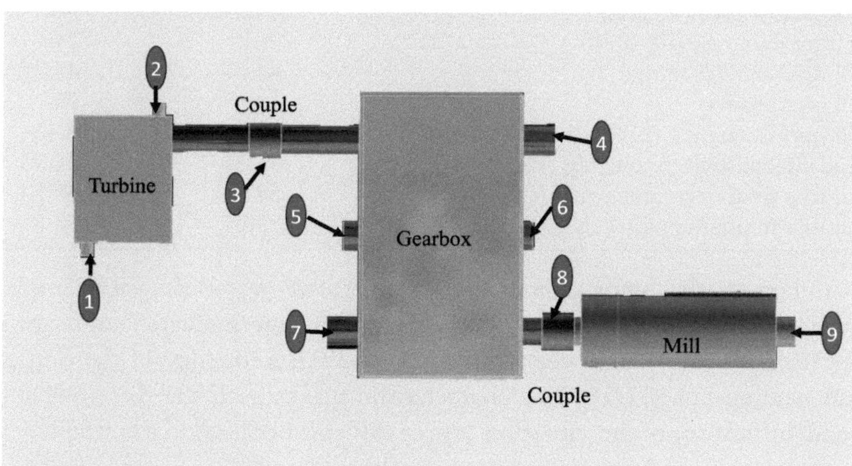


**Fig. 6.2** Diagram of the analyzed gear train.

## First reduction phase

The specific data for the first gear stage are presented in Table 6.1. It is important to note that the pinion is mounted on a journal bearing, therefore the vibration response shows a wide band of frequencies around 0.40–0.48X. In the table, the values are reported as empty.

| | Pinion | Gear |
|---|---|---|
| Speed (rpm) | 2976 | 960.6 |
| Teeth number | 51 | 158 |
| Type of support | Journal bearing | Roller bearing |
| Gear frequency (GMF) (Hz) | 2529.6 | 2529.6 |
| Bearing diameter (*D*) | Not applicable (NA) | 205 mm |
| Ball diameter (*d*) | NA | 44 mm |
| Ball number | NA | 14 |
| Ball spin frequency (BSF) | NA | 12.57 Hz |
| Fundamental train frequency (FRF) | NA | 35.57 Hz |
| Ball pass frequency inner race (OBPI) | NA | 88.01 Hz |
| Ball pass frequency outer race (OBPE) | NA | 136.1 Hz |

## Second reduction phase

| | Pinion | Gear |
|---|---|---|
| Speed (rpm) | 960.6 | 253.3 |
| Teeth number | 42 | 158 |
| Type of support | Bearing | Bearing |
| Gear frequency (FE) (Hz) | 672.43 | 672.43 |
| Bearing diameter (*D*) (mm) | 205 | 240 |
| Ball diameter (*d*) (mm) | 44 | 44 |
| Ball number | 14 | 28 |
| Ball spin frequency (FRB) (Hz) | 12.57 | 2.43 |
| Fundamental train frequency (FRP) (Hz) | 35.57 | 24.93 |
| Ball pass frequency inner race (OBPI) (Hz) | 88 | 68.24 |
| Ball pass frequency outer race (OBPE) (Hz) | 136.1 | 84.8 |

It can be seen that some operation frequencies are very close, for example, the second harmonic of the frequency FRB of the intermediate shaft bearing ($2 \times 12.57 = 25.14$ Hz) is very similar to the FRP frequency of the output shaft bearing (24.93 Hz). This characteristic makes both the diagnosis and the identification of the vibration source difficult because, in the frequency spectrum, this can be seen as a single peak.

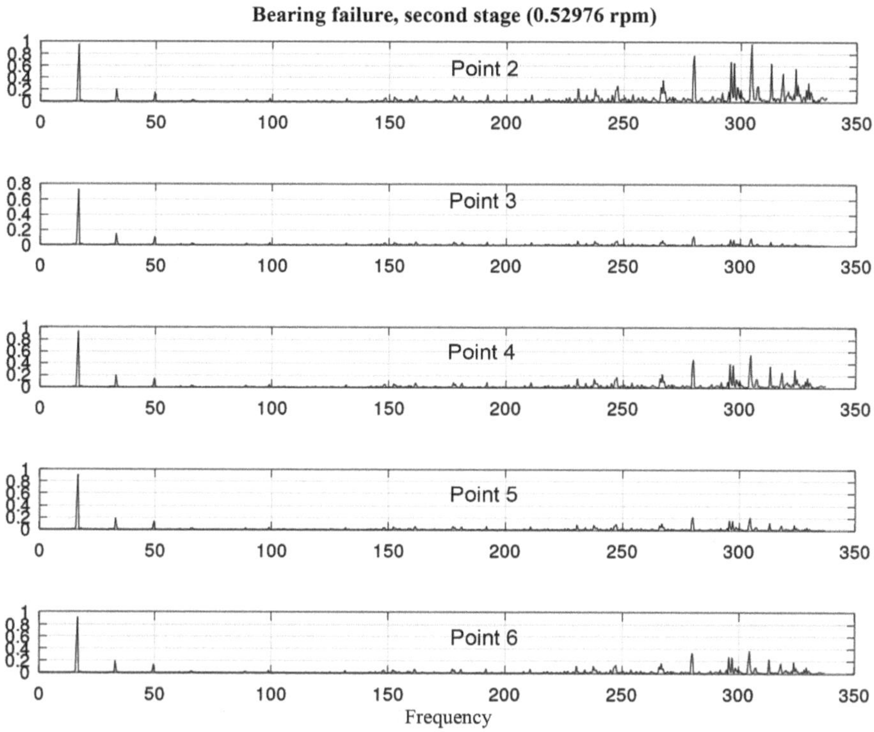


**Fig. 6.3** Frequency spectrum of the vibration taken in the indicated points in Fig. 6.2.

After defining the most important frequencies in the reducer, the vibration spectra (Fig. 6.3) were taken in all the indicated points in Fig. 6.2. In this case, the spectra that presented the highest peaks corresponded to points 5 and 6. Based on this information, point 5 was measured with more detail, registering the spectrum shown in Fig. 6.4. There, we can see the peaks corresponding to the ball spin frequency (FRB) and both the ball pass frequency outer race (OBPE) and the ball pass frequency inner race (OBPI) of the ball bearing from the pinion of the second speed reduction phase. In addition, in this point the increase in temperature was detected, making it possible to determine that the failure came from the ball bearing of the pinion, which was confirmed after making the corresponding repair.

## Gear in a speed reducer

In this case, a very significant increase was reported in the amplitude of vibration in the output shaft of a speed reducer, as can be seen schematically

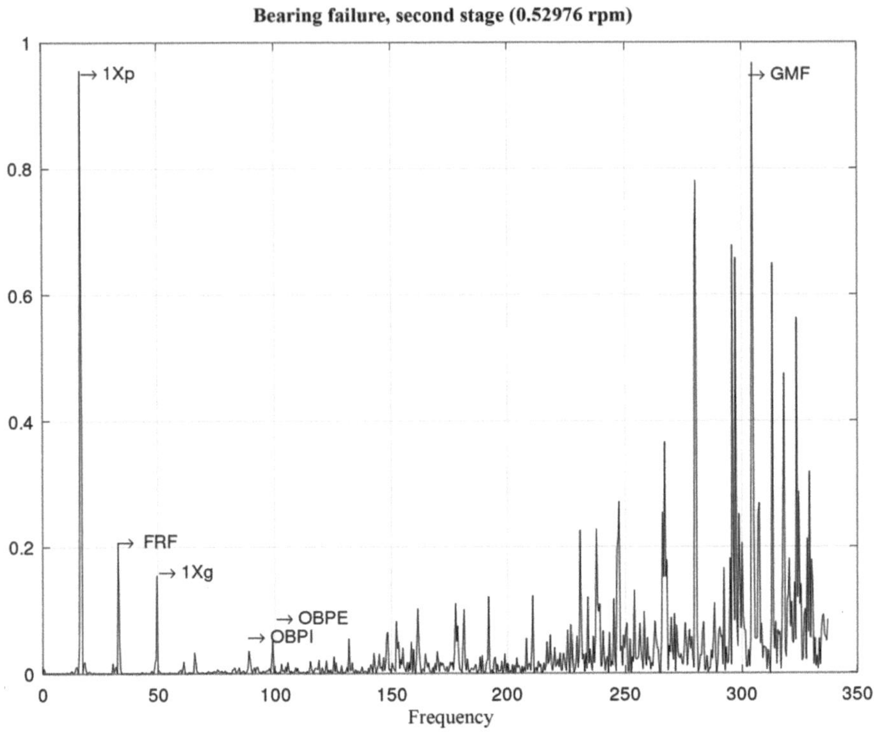


**Fig. 6.4** Frequency spectrum of point 5 of speed reducer.

in Fig. 6.5. This change in the amplitude is observed in the time domain. Fig. 6.6 shows a gear in good condition and gear with damage.

From this speed reducer, frequency spectra were taken from which an accelerated deterioration of the input pinion could be diagnosed; this can be verified directly from the spectrum shown in Fig. 6.7. For this diagnosis, it was essential to count with the vibratory signature of the gear because, based on this information, the interpretation of the frequency spectrum under abnormal operating conditions turned out to be relatively direct. That is, the gear frequency presents a very high peak as well as irregular sidebands.

## Excessive wear in gear teeth

The diagnosis corresponds to a rubber mill that produced rubber strips. The following case illustrates a typical vibration problem caused by excessive wear in the teeth of a gear. The problem was detected in the speed reducer of the rubber mill, which reported excessive thickness variations on the rubber

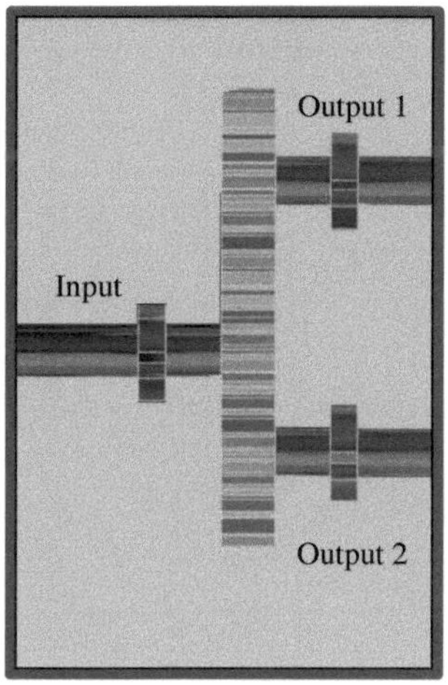


**Fig. 6.5** Schematic diagram of the speed reducer.

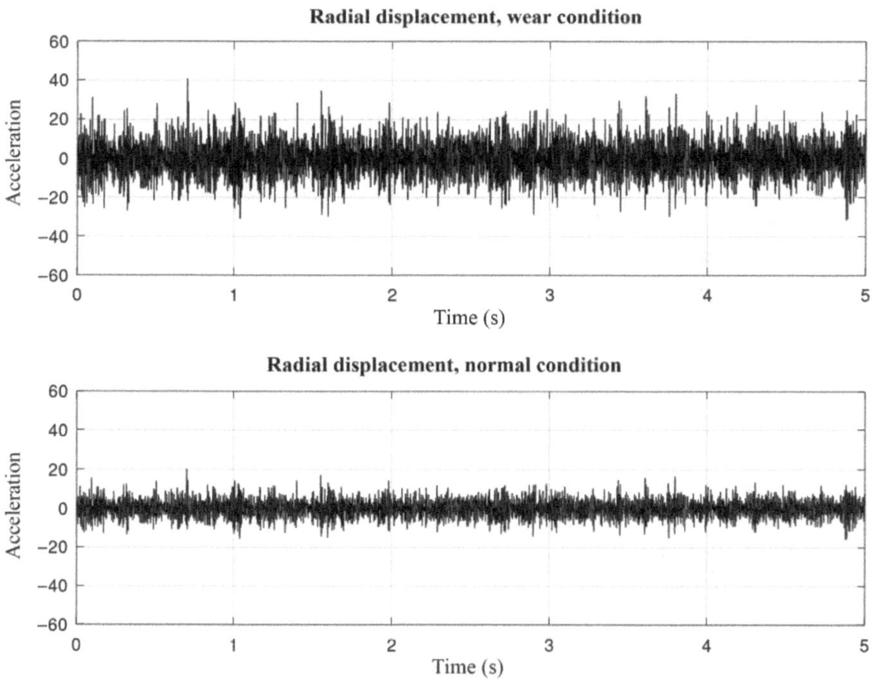


**Fig. 6.6** Vibratory signal of a gear operating normally and in fault conditions.

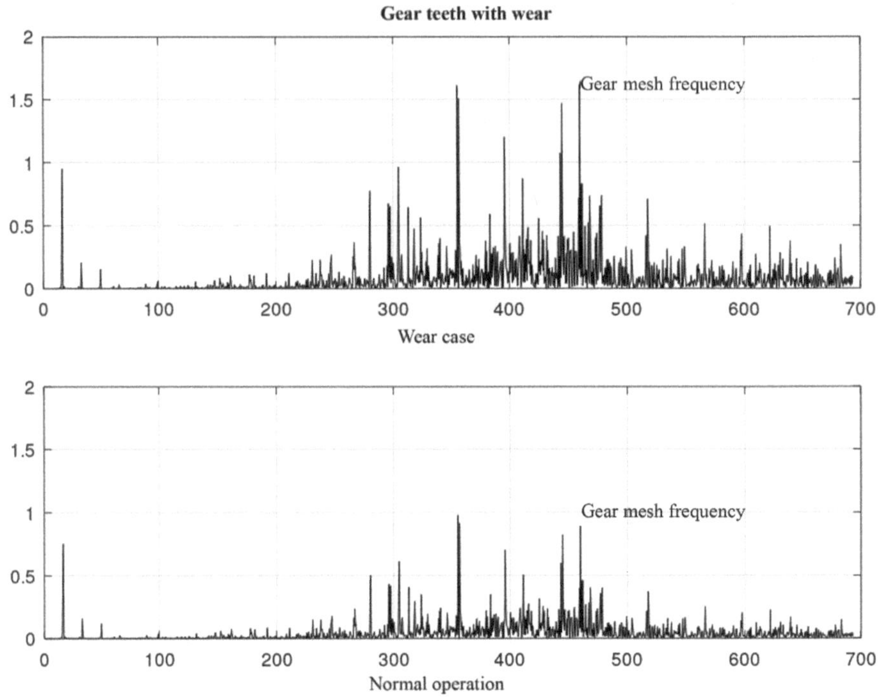


**Fig. 6.7** Frequency spectrum of a normal condition and a damage condition.

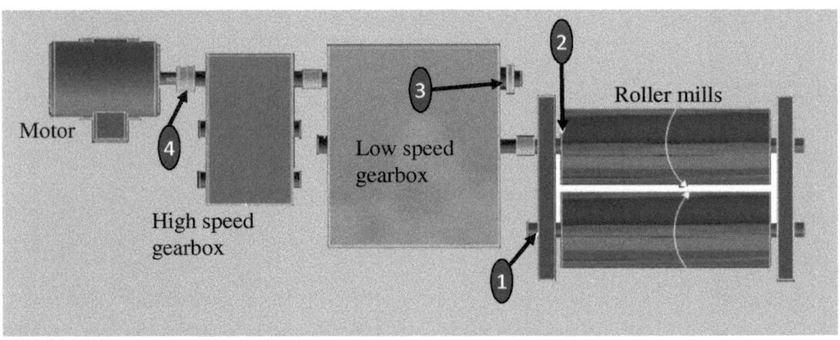


**Fig. 6.8** Schematic diagram of a roll mill, indicating the measuring points of vibrations for the purpose of an irregular movement diagnosis of the rollers.

strips, and it presented a seemingly random variation during the operation. In Fig. 6.8, it can be seen that the machinery train was formed by an electric motor, a first speed reducer of two stages, and a second reducer of only one stage that was directly connected to one of the rollers of the roll mill.

The report of the problem only specified the final effect, that is, that the rubber was rolled unevenly. The damage was identified by analyzing the frequency spectra in the indicated points (Fig. 6.8). At these points, three mutually perpendicular directions were acquired. The electric motor rotates at 1200 rpm, and the output shaft of the first reducer does it at 63 rpm. The rotation speed of the roller could not be registered with the accelerometer for being of a shallow frequency. Once these data were determined, all the spectra collected were analyzed. Those corresponding to point 1 were selected in the axial and transverse horizontal directions, in relation to the rotating shaft and the one measured in point 3 in the vertical transverse direction. These spectra are shown respectively in Fig. 6.9, where point 1 and point 3 can be seen, a prominent peak to the rotation frequency of the electric motor and several peaks at lower frequencies that correspond with the rotation speeds of the two-reducer shafts. A peak also sticks out at 1780 rpm, which agrees with the gear frequency (FE) of the second reducer. By identifying the harmonics of this gear frequency, it was noted that their second harmonic (3560 rpm) also had a prominent peak. By analyzing the spectrum taken in point 3 in the vertical transverse direction (Fig. 6.9), the corresponding peaks to the gear frequency were the highest, but above all, the presence of sidebands around the gear frequency and its harmonics was noted very clearly. This latter condition is the characteristic

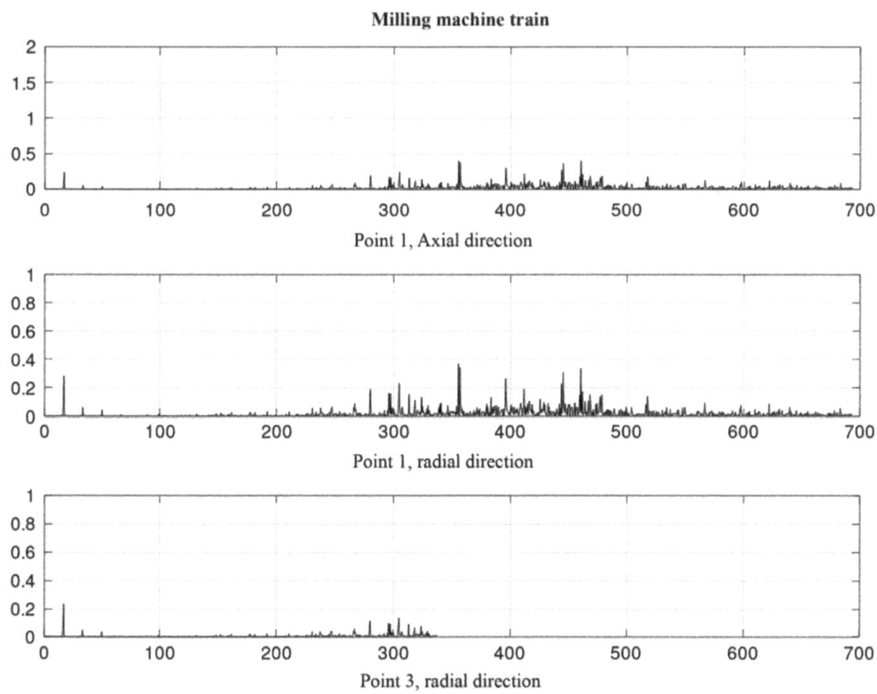


**Fig. 6.9** Frequency spectra of selected points in the gear train shown in Fig. 6.8.

that produces a gear drive with worn teeth because it does not have a combined action between two envelope curves (teeth in good condition), the vibration gets modulated by the rotation of the shaft (sidebands), and the shape of the wave presents different harmonics.

Along with the previous data, the possible cause of irregularities in the laminated rubber could be diagnosed as a gear with worn teeth because the detected pattern of vibration corresponded to this situation. Once the equipment was stopped, the reducer was uncovered, confirming the presented diagnosis.

## Machine tool

The following case corresponds to a high production drill. It was detected that the drill holes were out of perpendicularity, a condition that fostered a high number of rejected pieces because they came out of tolerance, affecting the subsequent assembly. After a visual inspection of the drill, performing tests without load, and inspecting the spindles, it was found that the machine was in good condition. Vibration analysis was carried out on the drill, and the measurements of the amplitude vibration were made at point 1, shown in Fig. 6.10. The measurements were taken with accelerometers, but the amplitude was set to display displacements. The spectra obtained did not explain any anomaly; however, when the amplitude was set to display acceleration, the spectra showed more information. Fig. 6.11 shows the corresponding frequencies in three machining directions ($x$, $y$, and $z$). $z$ corresponds to the axial displacement of the drill, and the spectra correspond to the displacement and acceleration when the drill operates fully loaded.

Fig. 6.12 shows the measurements with the amplitude in the acceleration scale. By comparing Figs. 6.11 and 6.12, it is noticed that the displacement

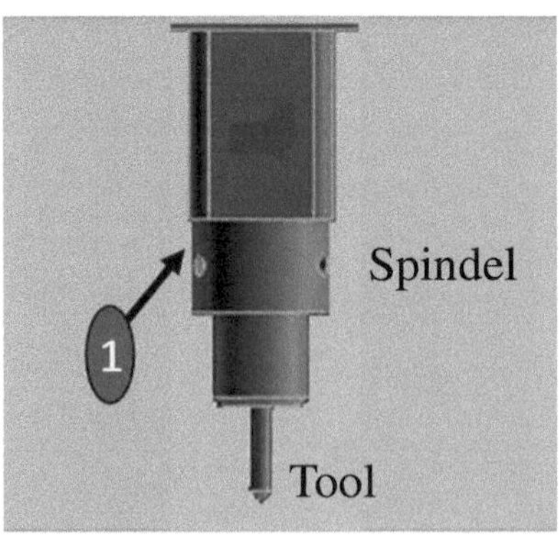


**Fig. 6.10** Location of the measuring point for obtaining the frequency spectra.

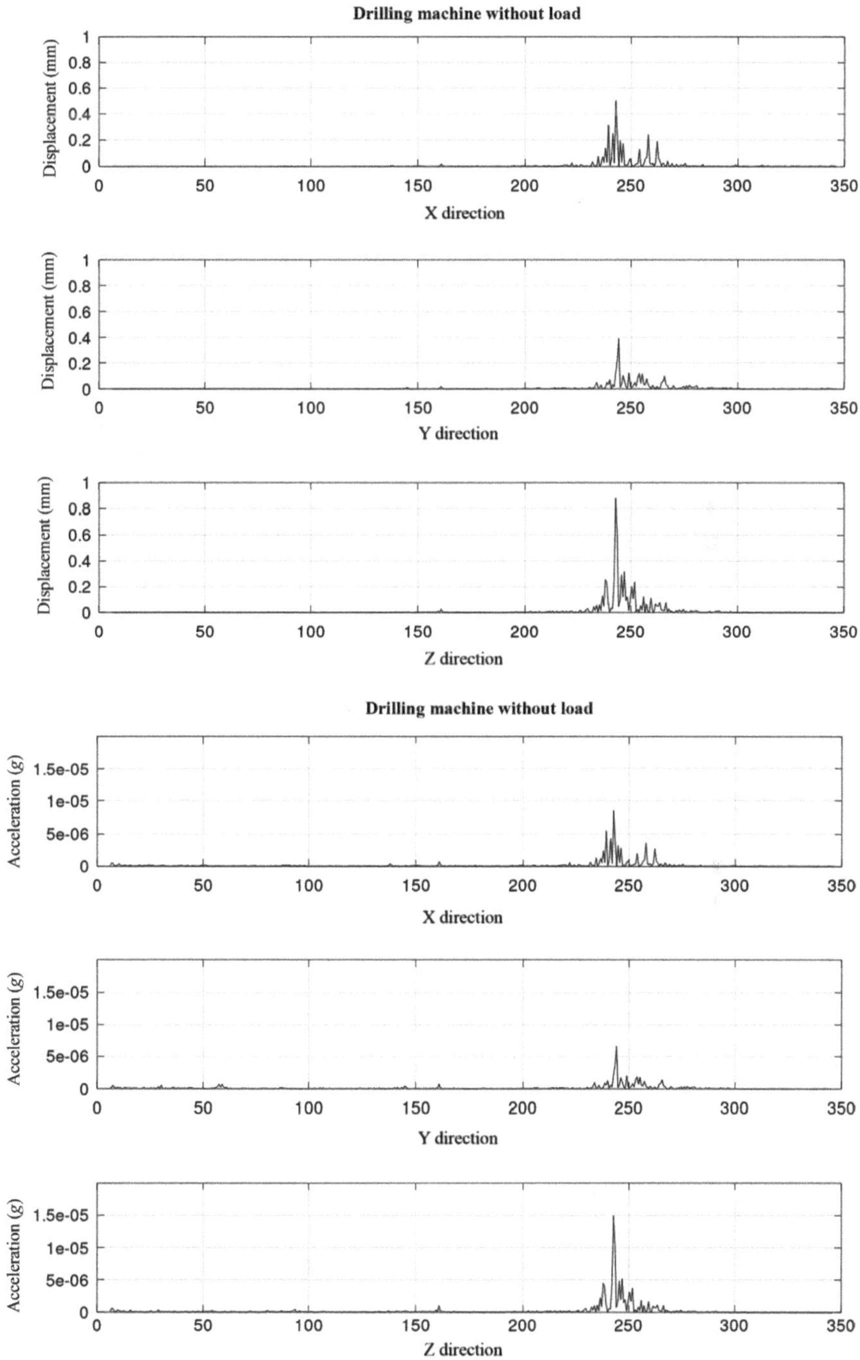


**Fig. 6.11** Frequency spectra for the measuring points without load.

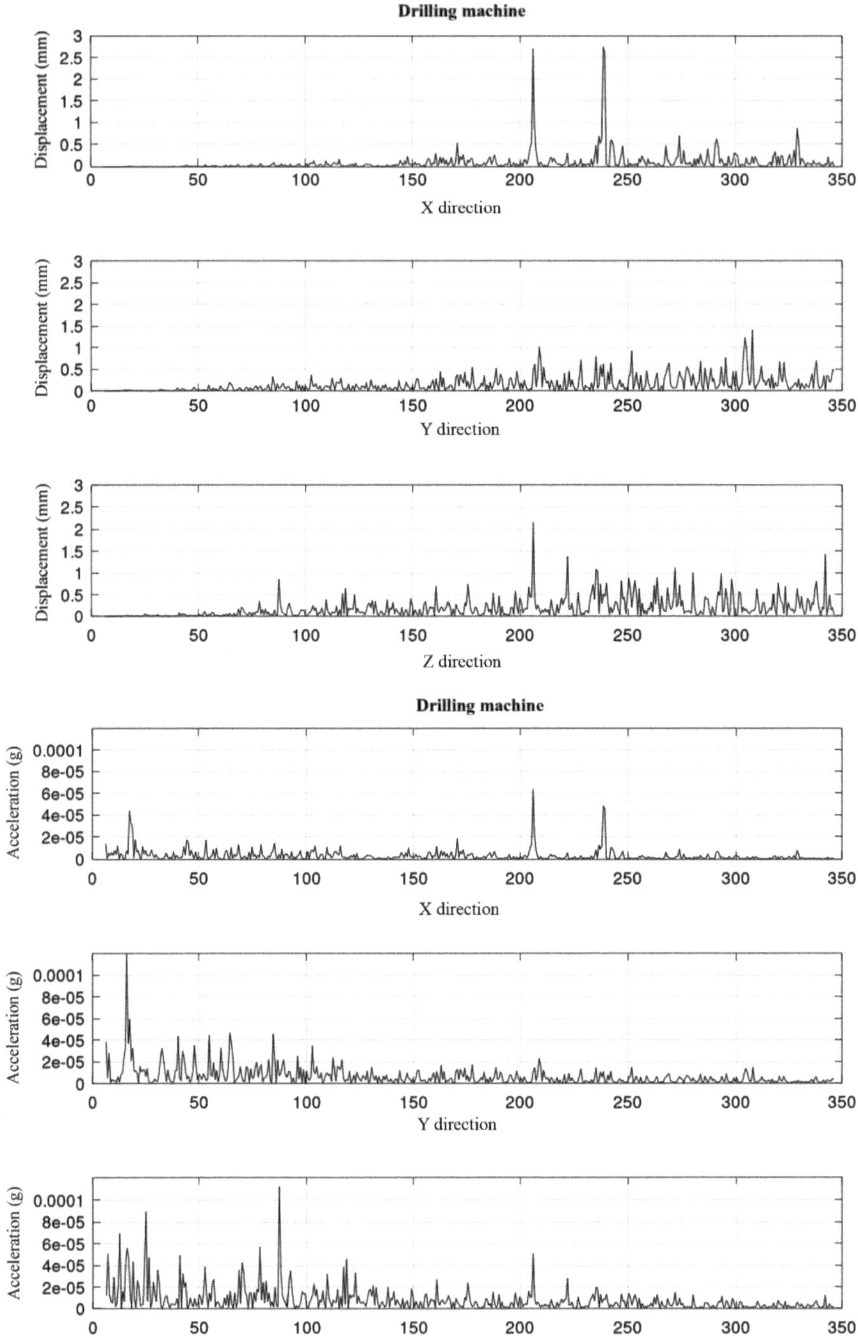


**Fig. 6.12** Frequency spectra for the measuring point under load (three directions).

amplitude only shows the high amplitudes around 230 Hz. The analysis of the unload condition indicates that the highest peak corresponds to the natural frequency.

With the measurements made in the acceleration scale, peaks in the frequency spectrum could be observed. In the $x$ and $y$ directions, the dominant peak occurred at 20 Hz, whereas in the $z$ direction it appears at 25 and 80 Hz. From these graphs, the difference of information provided by the spectrum with the use of one or another scale unit can be seen. The difference in these values identifies the origin of the machining defects; the peaks in the $x$ and $y$ direction are the cutting force while the peak in the $z$ direction corresponds to the bearings in the spindle.

Based on those mentioned above, the vibration levels are better identified using the acceleration scale. Fig. 6.12 shows the spectrum produced by the cutting force and the dominant frequencies that can be related to the failure problem. When the amplitude was set to displacement, it reduced the significant values, and the magnitude set to acceleration amplified the critical frequencies. Based on the measurements made, it was estimated that the problem came from the bearing because the dominant vibration corresponds to the bearing frequencies, regardless of the application of the cutting force. After changing the bearing, the machine returned to its normal operating conditions.

This example illustrates an important aspect; the proper selection of the scales of the spectra determines the effectiveness of the analysis. A poor choice of the scales to be used in the frequency spectra could hide the causes of vibration. It is also necessary to verify the operating conditions, which modify the measurements made, because the dynamic response of the equipment is different when operating with a load than without it.

## Bearing in an electric motor

The graph of supervision regarding the gear train vibration shown in Fig. 6.13 suggested a positive tendency, so it was therefore decided to carry out a detailed analysis of the vibration spectra. Those results are shown in Fig. 6.14.

From the previous figures, it could be identified that, in all spectra, there were vibration components with no apparent order because no peak corresponded to the angular speed of the gear train input and output, or to its harmonics. However, since the spectrum did not present an critical component at the operating speed, in this case at 1800 rpm, several possible causes were eliminated, such as imbalance, misalignment, and a shaft bent. The absence of problems with the gears due to the lack of sidebands in the

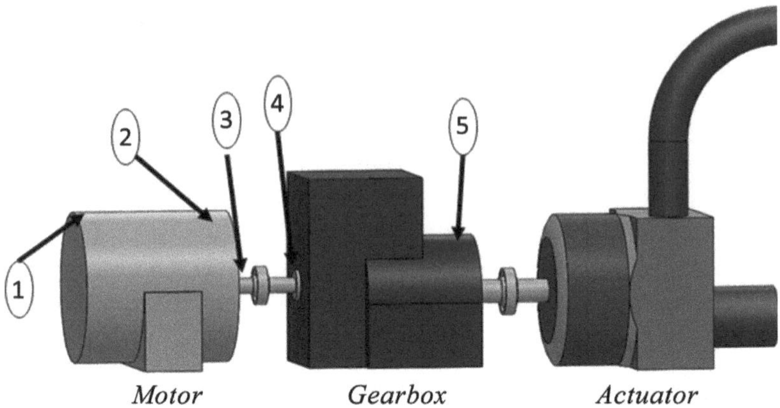


**Fig. 6.13** Measuring points of the gear train under study.

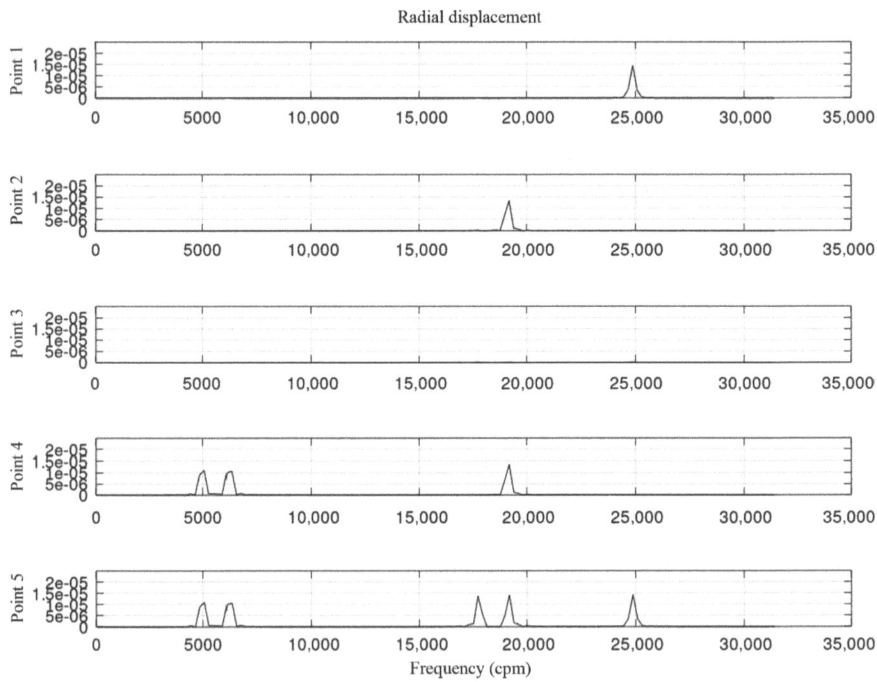


**Fig. 6.14** Frequency spectra for the indicated points in the gear train of Fig. 6.13.

spectrum was also estimated. In this way, the possible cause of high vibration in the gear train could be attributed to the bearings. To verify this assumption, an approach of the frequency spectrum of point 3 of the train was obtained, demonstrating that the frequency of the highest amplitude peak in the spectrum was in the range of the ball pass frequency outer race (OBPE) of the bearings used in this train. Based on the previous, it was suggested to change

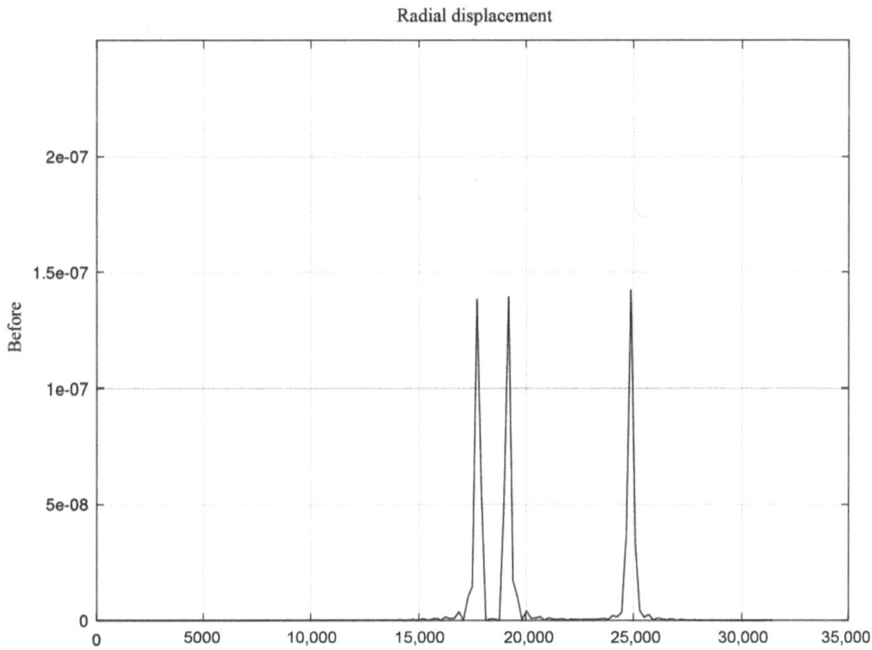


**Fig. 6.15** Frequency spectrum of point 3 of the gear train shown in Fig. 6.13, before and after the repair.

the bearing from point 3, obtaining the result shown in Fig. 6.15, which indicates the vibration levels in this point before and after the repair.

## Unbalance in a motor-fan assembly

In Fig. 6.16, a motor-fan assembly arrangement is shown. Despite balancing both the motor and the fan, problems continuously arose in this system due to high vibration in the fan and the necessity to repair the motor frequently because it burned out without visible cause.

The frequency spectra of the indicated points (Fig. 6.17) clearly show the operating frequency of the fan and the one of the motor, which suggests a problem of imbalance in the motor-fan coupling. To clear up this situation, the spectrum of points 1 and 2 of the motor (Fig. 6.18) was analyzed. It was observed that in the rotor of the motor, on the side where the pulley is located (point 2), the amplitude of vibration was greater than on the other end (point 1). This was due to the load of the band over the rotor of the motor, so it was necessary to balance it dynamically with the load of the present fan because vibration did not disappear after it was balanced without load and when it was put into operation.

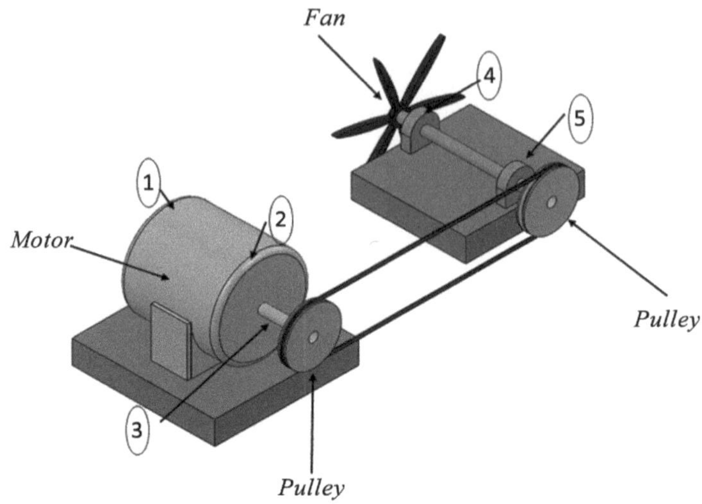


**Fig. 6.16** Motor-fan assembly and measuring points for the analyzed case.

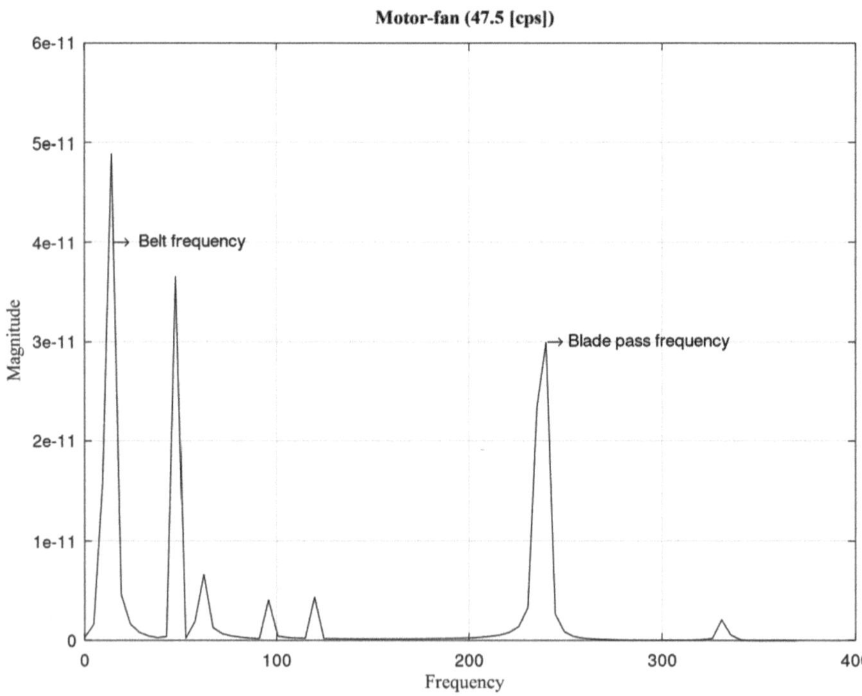


**Fig. 6.17** Frequency spectra of the indicated points in Fig. 6.16.

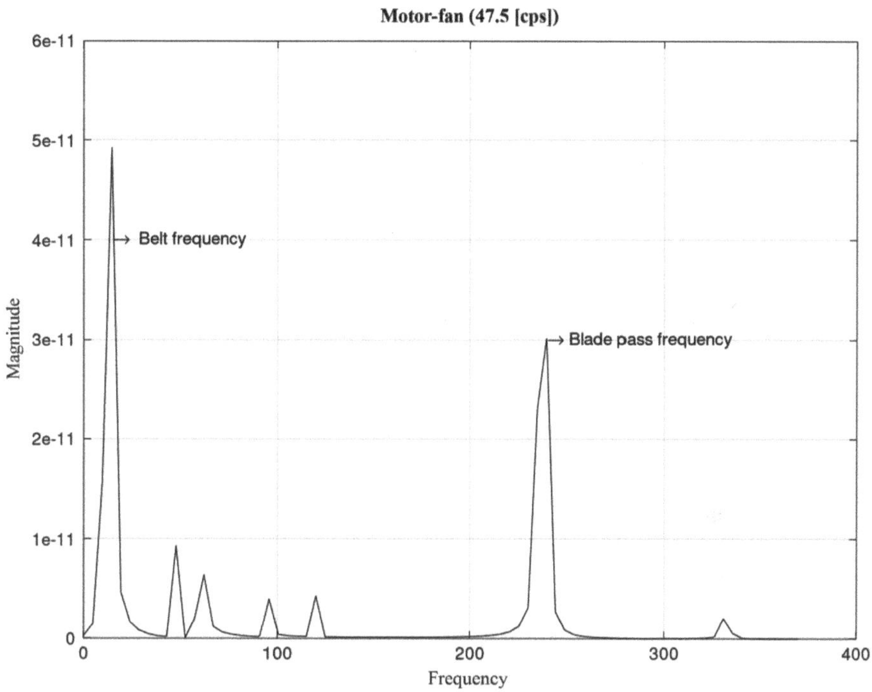


**Fig. 6.18** Detail of the frequency spectrum of points 1 and 2 of the motor-fan assembly of Fig. 6.16.

Fig. 6.18 shows that the peak at the rotation speed was reduced when the motor was balanced with the fan and the belt installed. The amplitude at the frequency of the belt was challenging to decrease because the tension remained the same.

## Centrifugal pump

In a water-pumping system, a leak in the discharge piping occurred. To get it repaired, a bypass was installed with a smaller diameter pipe. This, apparently, did not represent a problem in the functioning of the pumping system. However, shortly after, a very noticeable vibration appeared in the structure that supported the system, so it was decided to diagnose it. In Fig. 6.19, a schematic diagram of this installation is shown as well as the points where the frequency spectra were taken.

In each of the indicated points, the vibrations were measured in three orthogonal directions to record the possible sources of vibration. Fig. 6.20 shows the spectrum taken at position 3 in the horizontal direction.

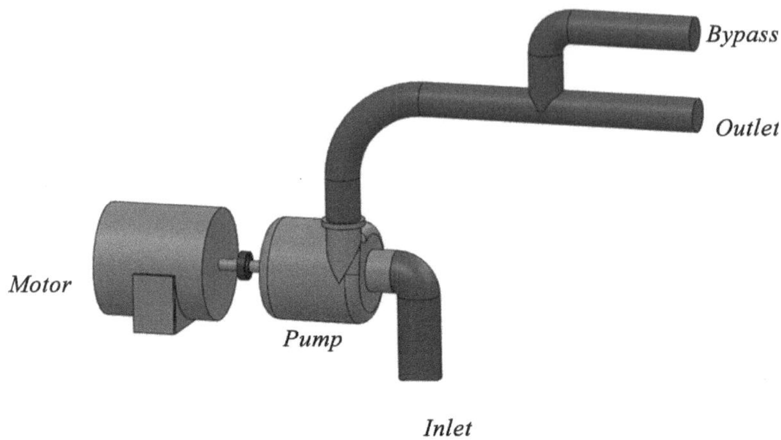


**Fig. 6.19** Scheme of the pumping system installation and the points where the vibration spectra were captured.

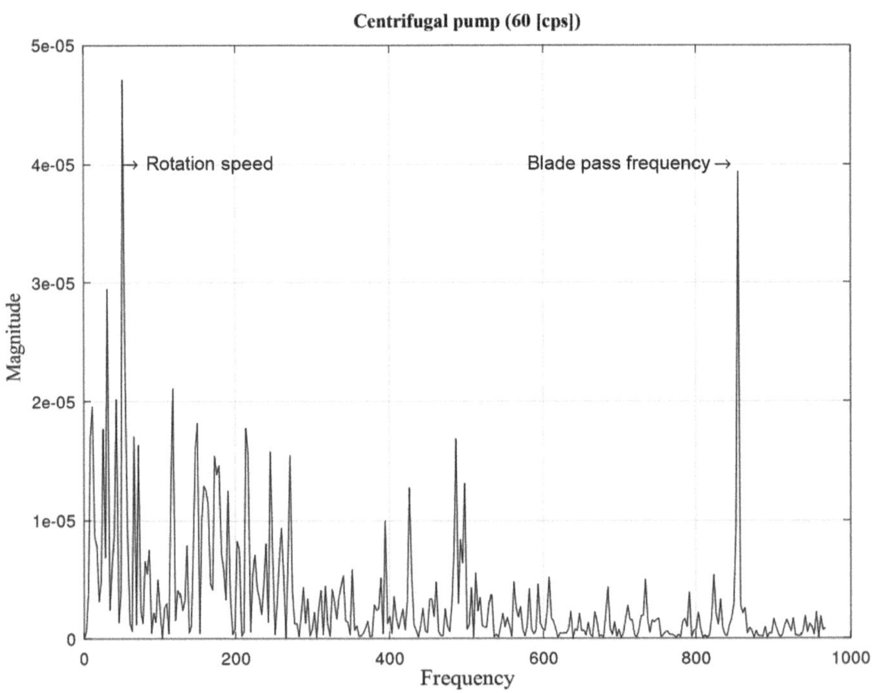


**Fig. 6.20** Vibration frequency spectrum taken in point 3.

This spectrum presented the highest amplitude at the rotating speed and the blade-pass frequency. Another characteristic of this spectrum was that other similar amplitude peaks suggested that a region of instability around these frequencies was occurring.

When analyzing the characteristics of the pump design, the source that produced this peak could not be identified. Therefore, to know its origin, another spectrum was taken in the structure that supports the piping (point 4), identifying this same peak as indicated by Fig. 6.21.

The spectrum that is shown in Fig. 6.20 suggested that the cause of vibration could be a natural frequency of the support structure. The electric motor was switched off, and the pump was left to stop gradually to test the hypothesis. A spectra cascade was captured, and it was observed that the peak of interest only changed in amplitude. The rest of the peaks lost amplitude and diminished their frequency as the rotation speed of the pump went down. In this way, it could be deduced that the vibration corresponded to a natural frequency of the structure. It was found that the modifications made in the pipeline produced turbulence, which was detected as instability

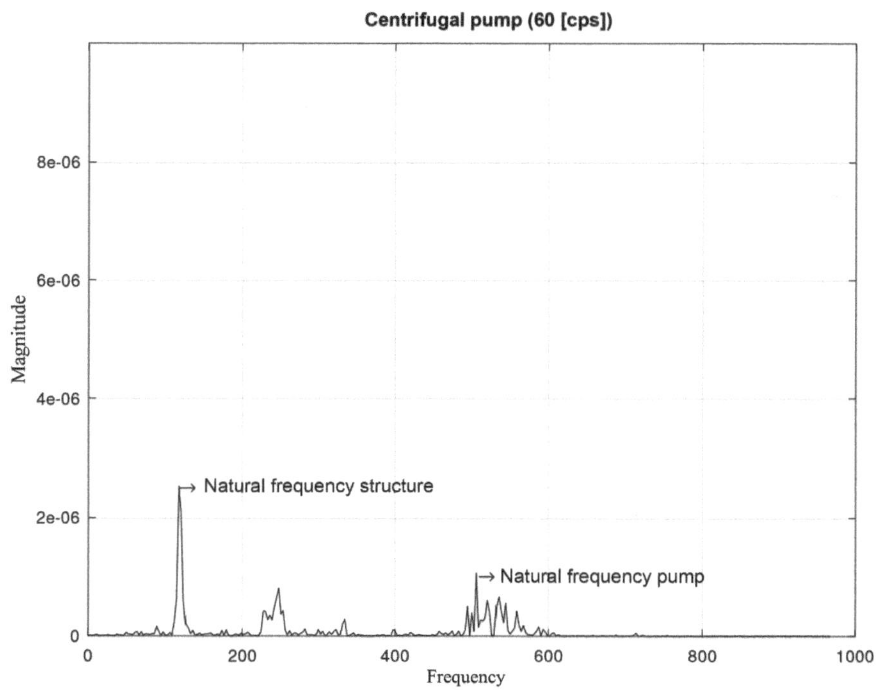


**Fig. 6.21** Vibration frequency spectrum taken in point 4.

in the frequency spectrum and coincided with a natural frequency of the structure. The problem was solved by adding extra supports to the new pipeline.

The case studies presented in this chapter reflect the challenges in setting a condition monitoring system. The identification of the dominant frequencies that appear in a machine train is the first step. Then, it is necessary to determine the normal variations of the amplitude measurements (a baseline), the nature of the dynamic response (linear or nonlinear), and the status of the entire components. The following chapter describes the philosophy behind a predictive maintenance system. This system is the preliminary component for the condition monitoring system, which elaborates the analysis techniques and the evaluation procedures for predicting failures at the earliest stage possible.

CHAPTER SEVEN

# Guidelines for the implementation of a predictive maintenance program

## Introduction

Conditioning monitoring systems evolved from predictive maintenance programs. The description of these programs will illustrate the definition of CMS and will provide the basis for their interpretation. This chapter remarks on the effects of a predictive maintenance program on productivity and product quality. It describes the steps to be taken to implement a predictive maintenance program. Maintenance has evolved through time; at the beginning, equipment malfunctions were repaired only at the moment they showed up (corrective maintenance). Afterward, the industry developed preventive maintenance systems that consist of repair actions based on statistical studies of the frequency of faults as well as recommendations from the manufacturer. These maintenance systems became expensive because only some of the repairs were needed. Then predictive maintenance systems were developed. They are based on periodic monitoring of the operating conditions of the machines and the application of techniques to estimate the expected life remaining in each component. These systems allow the scheduling of repairs and prevent unexpected failures.

Currently, it is convenient that all production systems based on continuous improvement techniques include, as a fundamental part of their scheme, predictive maintenance because it allows the total operation of the plant to be adjusted dynamically based on the actual operating conditions.

Predictive maintenance is a system that not only benefits the maintenance area, but also influences the improvement of productivity, the quality of the products, and the profitability of the investment. Therefore, it should not be taken only as a monitoring system but also as an integral part of the production strategy in a plant.

*Mechanical Vibrations and Condition Monitoring*
https://doi.org/10.1016/B978-0-12-819796-7.00007-X

Predictive maintenance consists of periodically monitoring (sample) the operating conditions of the machinery as well as analyzing the behavioral tendencies of each component. With this information, it is possible to estimate the time in which failures will occur. In this way, maximization of the repair intervals is achieved as well as minimizing the operating costs associated with premature or catastrophic failures and emergency stoppages of the machinery. To emphasize the concept, in this chapter a machine train is defined as three or more machine components that are coupled together, acting as one machine. The predictive maintenance systems are based on the different nondestructive techniques of analysis, among which the following stand out:

- *Vibration monitoring*: This technique consists of measuring the amplitude and vibration frequency of the critical components that form the machine train. This method was discussed extensively in the previous chapters.
- *Monitoring of the parameters of the process*: This is the main element of the predictive maintenance system. It consists of periodically recording the real values of all the parameters of the machine train (consumed power in electric motors, air pressure in pneumatic systems, hydraulic pressure, steam pressure, temperature, etc.). The purpose of this method is to relate variations in the measured variables with the probability of failure. The selection of the variables depends on the structure of the machine train and the type of individual components.
- *Thermography*: This method is based on measuring the temperature at different points of the machine train. This measuring provides related information to the mechanical behavior of each component because when a fail starts in any element of the machine train, usually the temperature increases. In this way, the changes in temperature can be associated with a machinery component with the possibility that it fails.
- *Oil analysis*: This technique is used as a complementary tool that analyzes the origin of the substances that contaminate the lubricants. The analysis helps to infer which of the components of the machine train is suffering premature wear or a more accelerated deterioration.

Among these techniques, vibration monitoring is the one that covers the most significant number of failure causes. Each of them occurs at a particular vibration frequency (as was seen in previous chapters) and their amplitude is a reflection of the machinery dynamic conditions. Besides, this amplitude in general increases according to the failure. The estimation of the time-to-failure stage defines the maintenance program. This estimation permits planning the entire production, meaning that the supply of spare parts can be done "just in time" and the maintenance activities can be determined ahead of time.

Predictive maintenance does not substitute for traditional maintenance systems; it is merely a tool that produces updated information allowing the evaluation of the future behavior of the equipment. The tendencies of all critical equipment can be graphed and estimated with a perfect approximation of the moment in which failure will occur. With vibration analysis, it is ensured that the repair intervals of the machinery will be extended and that the machine train offers the maximum availability of use. The benefits of predictive maintenance systems can be summarized as:

- Lower maintenance costs
- Lower number of failures
- Lower repair times
- Inventory reduction
- Increase the machinery lifespan
- increase in plant productivity
- Improvement of operational safety
- Verifications of operating conditions of new equipment
- Verification of repairs
- The possibility to implement "just in time" programs

Predictive maintenance programs start with the selection of the equipment to be monitored, the selection of the monitoring route, and the definition of the measuring intervals. Both the narrow bands and the alarm levels to each measuring point within the monitoring route are defined so that the tendency graphs are finally built, allowing the predictive dynamic behavior of the equipment.

## Definition of the monitoring route

For the implementation of the predictive maintenance program, all fundamental machines for the production process of the company must be selected. In this way, a database with the most relevant information and the operation efficiency is built.

In each machine, both the measuring points and the monitoring periods are identified. For each monitoring point, the total vibration spectrum is recorded (full firms), and the zooms around the characteristic frequencies are analyzed, in this way, it is possible to establish the starting point of the predictive maintenance program.

Subsequently, the monitoring intervals must be established to build the tendency curves and in this way schedule the repairs. When there are no initial reference points, it is therefore recommended to monitor each selected point at intervals of between 8 and 15 days. Once the data for the first five points are established, the most suitable inspection intervals can be determined with the objective of "measuring the vibrations at lower time intervals than the intervals in which variations are detected in the measured signal in normal operating conditions." This procedure minimizes the possibility of a catastrophic failure.

Fig. 7.1 presents the curve of failure frequencies of pieces of equipment from their installation until removal from production. This graph is known as the curve of life because it resembles human life. It defines three phases: youth, maturity, and old age.

- Youth
- Productive life
- Elder

In the youth phase, the failures are greater due to their initial adjustments, premature wear, settlements, or design changes. The mature stage is considered to be the productive lifespan of the equipment. In the old age stage, the failures increase due to the deterioration of all the systems.

The objective of predictive maintenance is to get the machine train to operate as long as possible in its production life stage at the lowest possible cost. In this way, it would be ideal that the machine be substituted out due to a significant change in technology and not because it has reached its old age.

For the correct identification of the measuring points, it is essential to know the design of the machine train as well as the functioning details of the components, such as the rotation speeds of each shaft that form the gear, the type of supports (ball bearings, bearings), the type of power equipment (turbine, electrical motor, internal combustion motor, etc.), and their operational parameters.

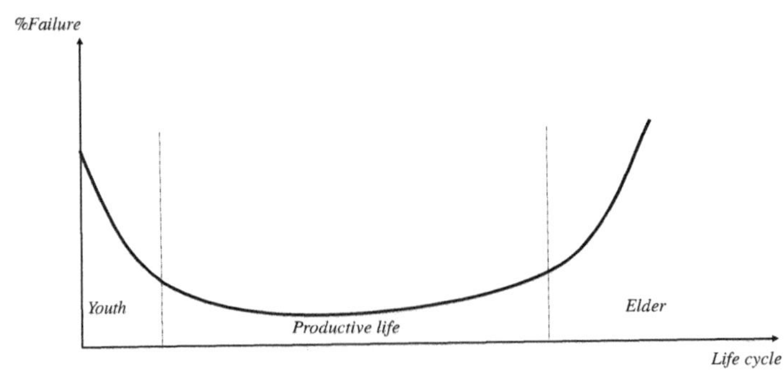


**Fig. 7.1** Life curve of a machine train.

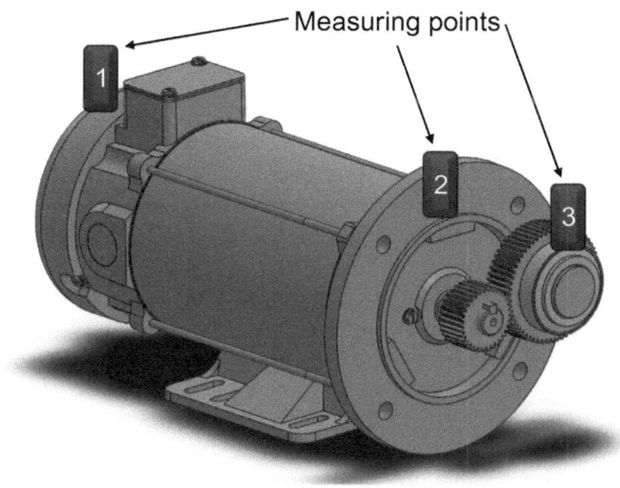


**Fig. 7.2** Components of a machine train and the location of monitoring points.

Fig. 7.2 is a schematic representation of a typical machine train. The numbering of the points where the vibratory firms are measured every sampling time is indicated. It is recommended to carry out the numbering sequence from the end support of the motor and culminate with the end support of the output shaft. In this way, there is an ordered sequence of points and information can be handled with more detail. In each point, the direction of the signal must be identified. For example, in Fig. 7.2, point 1 would allow identifying the level of vibration of the train in three directions: horizontal, vertical, and radial. Once the route is established, at each monitoring point the narrow bands are defined.

## Narrow band selection

Once the monitoring points in the gear train are located, the most outstanding characteristic frequencies of the vibration sources are identified. Around each of these frequencies, it is required to select the narrow bands to monitor before setting both the alert and alarm limits. These bands must contain the characteristic frequency as well as covering variations in the operating speed that are generally between 10% and 20% of the characteristic frequency when the speed variations of the equipment are small.

As a general guide, in Table 7.1 the frequencies are indicated where more information from the vibratory signal can be obtained, depending on the type of failure under study.

**Table 7.1** Frequencies and the sideband limits for different failure modes.

| Type of failure | Characteristic frequency (×) |
|---|---|
| Imbalance | Rotation frequency of each of the shafts with a bandwidth of 10% in the radial direction |
| Misalignment | Rotation frequency and its first harmonic with a bandwidth of 10% in both the radial and axial directions |
| Bent shaft | Rotation frequency and its first harmonic with a bandwidth of 10% in both the radial and axial directions |
| Eccentric shaft | Rotation frequency and its first harmonic with a bandwidth of 10% in both the radial and axial directions |
| Mechanical clearance | Rotation frequency of each shaft and to 0.5 ×, 1.5 ×, and first harmonic (2 ×) in the radial direction |
| Journal bearings | Rotation frequency and its first harmonic with a bandwidth of 10% in the radial directions<br>In addition to a narrow band between 0.42 × and 0.5 × in the radial direction |
| Belts | Rotation frequency of each of the shafts with a bandwidth of 10% in the radial direction<br>Pass frequency of the band and its first harmonic with a bandwidth of 10% in the radial direction |
| Electrical defects | Rotation frequency with a bandwidth of 10% in the radial direction<br>Pole pass frequency<br>Rotor bar pass frequency including at least four sidebands per side<br>Line frequency including at least four sidebands of the sliding frequency and its first harmonic |
| Hydraulic and aerodynamic | Rotation frequency with a bandwidth of 10% in the radial direction<br>Blade pass frequency and at least four sidebands<br>Turbulence (from 50 to 2000 cycles per minute, randomly)<br>Cavitation (In the case of the pumps, a signal appears between 0 and 1 ×, randomly.) |
| Roller bearings | Inner race pass frequency with a bandwidth of 20%<br>Outer race pass frequency with a bandwidth of 20%<br>Ball pass frequency with a bandwidth of 20%<br>Rolling-element pass frequency with a bandwidth of 20% |
| Gears | Rotation frequency of each of the shafts with a bandwidth of 10% in the radial direction<br>Gear frequency including at least four sidebands and the first harmonic of the gear frequency |

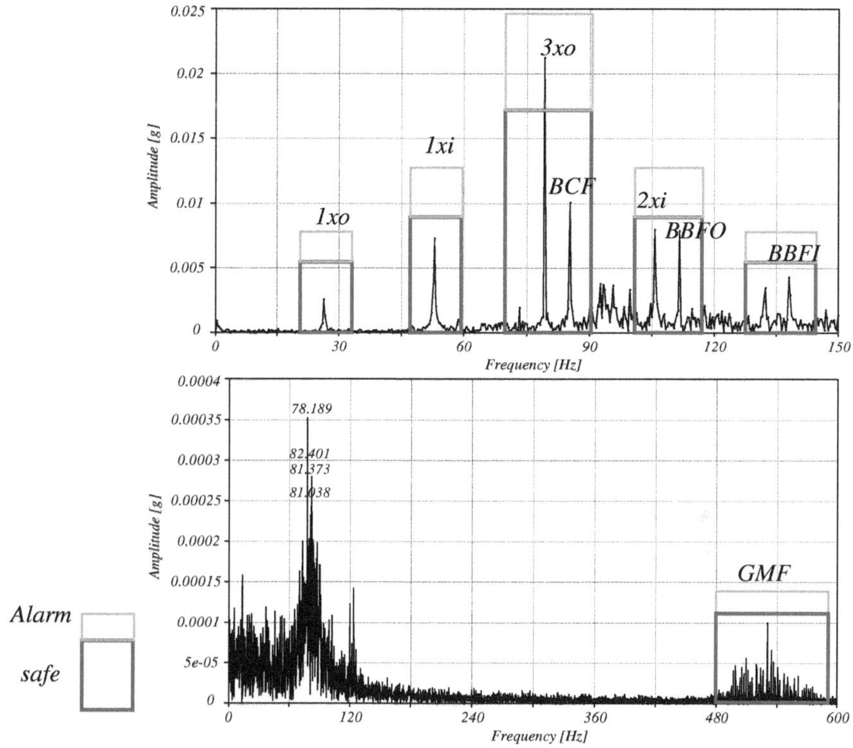


**Fig. 7.3** Example of a frequency spectrum with the location of the narrow bands.

Once the narrow bands are determined, the predictive maintenance program starts, analyzing the tendencies of the dynamic behavior of the selected gear train. In Fig. 7.3, a spectrum with its narrow bands is shown.

## Analysis of tendencies

Once the narrow bands in each monitoring point of the gear train under study are established, the permissible limits of the vibration levels are defined. The selection of these limits depends on the behavior of each machine, its importance in the production system, and the experience of the maintenance staff. As a general rule, it is recommended to establish the alert limit of measurements to twice the average value of the vibration amplitude in normal operating conditions, and the alarm limit to five times this latter value.

In general, the behavior of the vibration follows an exponential law. Therefore, when the vibration level reaches the alert level, the tendency curve can be traced and a date in which the vibration would reach the alarm limit can be objectively determined. During this period, repair or removing the machine from production must be done (Fig. 7.4).

The periodic monitoring of the vibration allows the stage of the life curve in which the gear train is located to be identified. The problem of tendency analysis is that it does not allow establishing the causes that produce the deterioration of the equipment. For that reason, it is necessary to complement the analysis with other information such as the reports of the operating conditions, the thermal analysis, the oil analysis and knowledge of the equipment design. This type of analysis complements the understanding of the equipment behavior, making the predictive maintenance more efficient.

Because the vibration in a machine does not keep the same levels of amplitude through its lifespan, it is necessary to provide a statistical followup to the behavior of the frequency under study to properly establish both the alert and alarm limits. Among the causes that produce the variation of the vibration amplitude, the changes in the operation of the machinery, climate changes, and the deterioration of the equipment itself stand out.

The analysis of the tendencies of the dynamic behavior of the machinery is the basis of the predictive maintenance system. To best illustrate its application, it is necessary to consider the subsequent example.

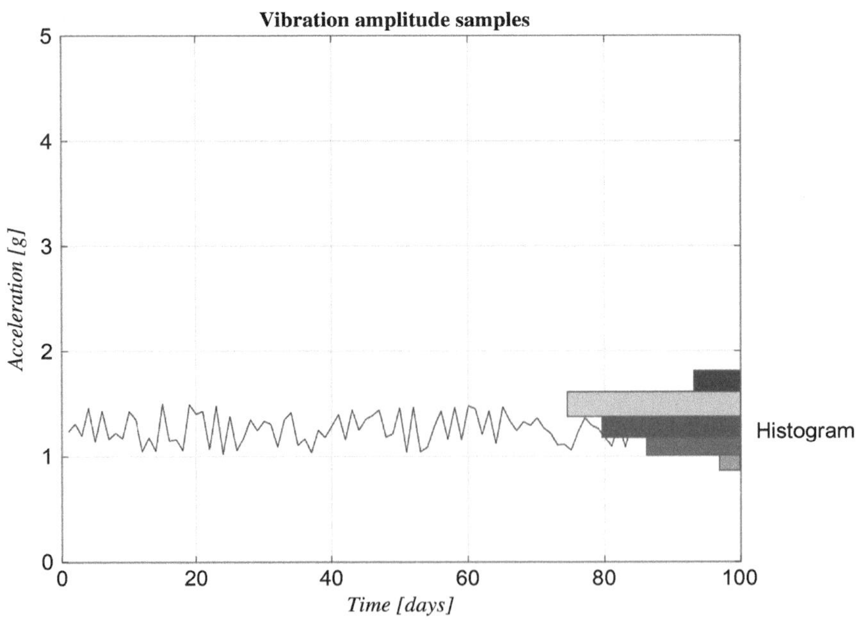


**Fig. 7.4** Analysis of tendencies and histogram.

## Example

In a chemical plant, there are two turbo-generators identical to the one shown in Fig. 7.5. The current importance of such equipment requires including them in the productive process. They were included in the predictive maintenance program of the entire installation. The turbo-generator can be defined as a machine train consisting of:

- A two-stage steam turbine with 135 and 150 blades, respectively, with a centrifugal governor for speed control.
- A gear reducer of a single-stage double-helical type with 21 teeth in the pinion and 97 in the gear, lubricated with a gear pump of eight teeth connected to the pinion shaft mounted on hydrodynamic ball bearings; the gear is mounted on bearings of 12 balls with diameters of ball pass $D=235$ mm and $d=40$ mm for each ball.
- Electric generator of six poles with a synchronous speed of 1200 rpm. The rotor is mounted on journal bearings.

Once the design is identified, the characteristics of all the components and the monitoring route are defined.

In this example, the dominant frequencies that each element generates are determined next.

## Monitoring route

For the example, four monitoring points were selected. The first was at the free end of the turbine and the latter at the free end of the generator.

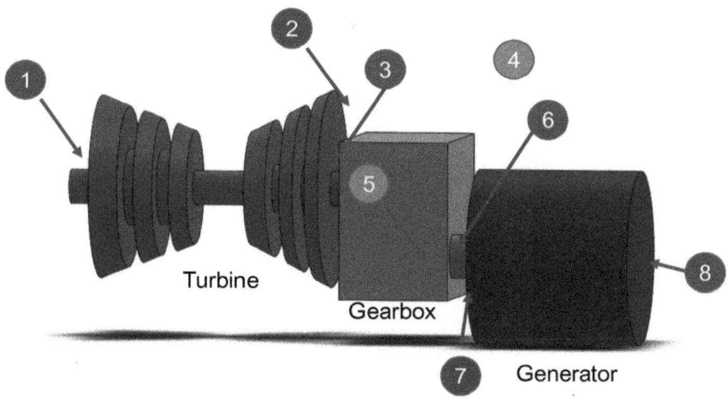


**Fig. 7.5** Schematic representation of a turbo-generator.

In each measuring point, the frequency ranges were determined in which spectra were recorded as well as the frequencies of interest. For this, the frequency values that characterize each component of the turbo-generator were calculated. In this case, the rotation frequency of the turbine shaft (point 1) was calculated, depending on the gear reduction relation, in such a way that:

$$\omega_T = \left(\frac{n_c}{n_p}\right)\omega_G$$
$$= \frac{97}{21}(1200) = 5542.8\ cpm$$

The blade pass frequency was calculated as:

$$FP_{A1} = N_{A1}\omega_T$$
$$= (135)5542.8$$
$$= 748{,}285.7\ cpm$$

And for the second pass:

$$FP_{A2} = N_{A2}\omega_T$$
$$= (150)5542.8$$
$$= 831{,}428.5\ cpm$$

For point 2, the rotation frequency of the pinion is the same as that of the turbine, so in these points, the frequencies of interest were those of the gear, the one of the teeth pass of the lubrication pump, and those associated with the problems of the hydrodynamic ball bearing.

The gear frequency is:

$$FE = n_p\omega_T$$
$$= (21)5542.8$$
$$= 116{,}398.8\ cpm$$

For the lubrication pump:

$$FE_B = n_B\omega_T$$
$$= (8)5542.8$$
$$= 44{,}342.4\ cpm$$

And for the ball bearings, a range between 0.35 and 0.5 was estimated, and Fch was:

$$Fch = 1940 \text{ to } 2770\ cpm$$

In point 4, the vibration was measured at the rotation frequency of the gear (8w = 1200 rpm) and to the frequencies that characterize the bearings:

$$FRB = \left({}^{1}/_{2}\right)\left(1 - r^{2}\right)x$$
$$FRP = \left({}^{1}/_{2}\right)(1 - r)x$$
$$OBPE = \left({}^{n}/_{2}\right)(1 - r)x$$
$$OBPI = \left({}^{n}/_{2}\right)(1 + r)x$$

With the data of the ball bearings, the fundamental frequency of the bearing (fundamental train frequency) was determined, having:

$$FRP = 497.87\ cpm$$

The ball spin frequency is:

$$FRB = 3422.8\ cpm$$

The frequency due to a fault in the outer race is:

$$OBPE = 59744.4\ cpm$$

And the frequency due to a failure in the inner race is:

$$OBPI = 8425.53\ cpm$$

In the measuring points that correspond to the electrical generator, both the power line frequency and its harmonics (in this case, there is no slip frequency) were specified, in addition to the rotation frequency.

$$F_L = 3600\ cpm$$
$$2F_L = 7200\ cpm$$

Once the characteristic frequencies and the narrow bands for each of the selected measuring points was identified, the narrow bands were selected based on the previous information in this chapter, rounding their values according to the resolution of the used vibration analyzer equipment (Table 7.2).

### *Analysis of tendencies*

Once both the monitoring route and narrow bands were established, it will proceed to the capture of the frequency spectra. At the beginning of the program, it is required to monitor the system under study with higher frequency. This frequency depends on the type and operating condition of the gear train. However, as an average value, it is estimated that by carrying out

**Table 7.2** Characteristic frequencies and narrow bands of the measuring points selected.

| Measuring point | Direction | Reference frequency (cpm) | Side bands |
|---|---|---|---|
| R01 | Radial | 5500 | 5000–5750 |
| | | 11,100 | 10,000–12,200 |
| Z01 | Zoom (radial) | 750,000 | 725,000–770,000 |
| R02 | Radial | 5500 | 5000–5750 |
| | | 11,100 | 10,000–12,200 |
| Z02 | Zoom (axial) | 830,000 | 810,000–855,000 |
| R03 | Radial | 2300 | 1940–2780 |
| | | 5550 | 5000–5750 |
| | | 11,000 | 10,000–12,000 |
| A03 | Axial | 5500 | 5000–5750 |
| | | 11,100 | 10,000–12,200 |
| Z03 | Zoom | 116,000 | 93,800–138,000 |
| R04 | Radial | 2300 | 1940–2780 |
| | | 5500 | 5000–5750 |
| | | 11,000 | 10,000–12,000 |
| A04 | Axial | 5500 | 5000–5750 |
| | | 11,100 | 10,000–12,200 |
| R05 | Radial | 500 | 400–600 |
| | | 1200 | 1100–1320 |
| | | 2400 | 2200–2600 |
| | | 3400 | 3000–3500 |
| | | 6000 | 5750–6600 |
| A05 | Axial | 1200 | 1100–1320 |
| | | 2400 | 2200–2600 |
| R06 | Radial | 500 | 400–600 |
| | | 1200 | 1100–1320 |
| | | 2400 | 2200–2600 |
| | | 3400 | 3000–3500 |
| | | 6000 | 5750–6600 |
| A06 | Axial | 1200 | 1100–1320 |
| | | 2400 | 2200–2600 |
| R07 | Radial | 1200 | 1100–1320 |
| | | 2400 | 2200–2600 |
| | | 3600 | 3550–3650 |
| | | 7200 | 7150–7250 |
| A07 | Axial | 1200 | 1100–1320 |
| | | 2400 | 2200–2600 |
| R08 | Radial | 1200 | 1100–1320 |
| | | 2400 | 2200–2600 |
| | | 3600 | 3550–3650 |
| | | 7200 | 7150–7250 |

measurements every 15 days, a pattern of behavior of the mechanical rotating equipment can be established. With the information collected, the analyst graphs the tendency curves for each point and adjusts the reading intervals according to the equipment performance.

Even though a large amount of data is handled, modern computer systems allow its analysis in real time, which facilitates its handling, allowing the maintenance engineer to define both the inspection intervals and the repair of the machinery with objective criteria.

CHAPTER EIGHT

# Condition monitoring

## Introduction

Predicting life has been a challenge for every generation. It is the core question for most scientific research programs and technological developments. This thinking backs up the integration of different available technologies for the diagnosis and estimation of the remaining life of any system. If we map the evolution of modern society, it is clear that tools and machines are better and more sophisticated year after year. And, as devices become more complex, the need for a more profound knowledge of their behavior as well as the necessity of accurate data become more demanding. Predicting life today combines the ability to understand physics and detect physical variables as well as computerized algorithms to process the data to be able to detect a possible failure or miss function. These concepts are the fundamental philosophy of any condition monitoring system. The basic idea of predicting faults in machinery is the analysis of the dynamic response of a mechanical system. A machine can be represented with a lumped-mass-model. Assuming that the model is strongly linear, the output response will be proportional to the input signal. In practice, the input signal will be related to the excitation forces of the mechanical components that constitute the machine, and the output response will be the displacements of the supports. This assumption is valid for most rotating industrial machines, and those machines that have a nonlinear response will have a specific output response. The evolution of sensors made possible measuring the dynamic response with an accelerometer. There are different technologies such as seismic and piezoelectric of capacitive (MEMs) transducers. These transducers detect vibrations in specific directions, and their measurements are the actual output response.

Condition monitoring is an evolution of predictive maintenance or proactive maintenance. The origin is difficult to define, but predictive maintenance has made enormous progress over the last few decades. Nowadays, it has been addressed as one of the most innovative solutions for anticipating failures in machinery and is being used by a wide variety of industrial sectors.

*Mechanical Vibrations and Condition Monitoring*
https://doi.org/10.1016/B978-0-12-819796-7.00008-1

Prediction maintenance could be applied to a large industrial sector when the cost of vibration sensors is competitive. This advantage reduces the cost of failures in comparison with the investment in the measurement equipment and analysis system. In the beginning, the systems were rudimentary and required specialized personnel to collect data and make the analysis. The convergence of high-precision accelerometers and the ability to process the Fourier transform with the FFT algorithm allowed the development of rapid tools that can make a diagnosis on the actual condition of machines. Previously, vibration sensors were applied in a few types of equipment due to their cost and the need for specialized personnel. These first concepts were complemented with other emerging technologies such as ultrasonic, thermography, acoustic sensors, and directional microphones.

The first application of predictive maintenance was made at the Royal Air Force in the United Kingdom. It was found that the rate of failure increased after the repair or inspection of machines, even following the maintenance plans. This phenomenon was named the "Waddington Effect," which led to condition monitoring. It was decided to adjust the maintenance programs and align them to the physical condition and frequency of use to reduce the Waddington effect. The process required the analysis of much data, but the launch of this program reduced the number of failures. Condition monitoring systems (CMS) evaluate the vibration data and determine the condition of the machine based on the analysis of amplitude and frequency. The original signal has raw data that have to be treated to produce a reference baseline. Sampling the evolution of the data during operation indicates the condition of the machine and, in the case of a failure, the data will present significant changes. Condition monitoring systems have increased the reliability of machinery because they include new sensors; they use fast processing hardware and better algorithms for the signal process. The application of artificial intelligence programs in condition monitoring systems has increased the reliability of modern machinery, allowing more extended periods between maintenance. These complex systems can anticipate failures in most of the components that constitute modern machines. The analysis data is also linked to the purchasing programs and supply chain, enabling the reduction of spare part inventories.

These systems aren't limited to increasing the reliability of equipment; they are also the source of data for the modern concept of "twin machines," which are computer replicas of real machines operating in industry or airplanes. The computer models are continuously adjusted with the information recorded through the sensors and analysis systems that constitute the condition monitoring system.

## Description

Condition monitoring systems are crucial for understanding the behavior of machinery, in particular rotating machinery. They are a key element for predicting failures in most of the elements that compound a machine. They are able to identify the type of fault and the instant when a defect appears in a particular component. They also provide enough data for improving machinery design that can be analyzed to correlate the operating loads, the life cycle, and the type of failure. These data are an important input for future designs. The construction and definition of a condition monitoring system defines the type and amount of data that will be generated once the machine is set and operates.

The design of a condition monitoring system must be robust and reliable because it must determine the instant when a failure appears. This important feature has backed up most of the studies around CMS that are focused on sensors, analysis techniques, predicting algorithms, and different measurement devices. Among the measuring systems, vibration measurement is still the most reliable technique for defining a CMS, and it can be improved by complementing the data with measurements obtained with other instruments. The drawback of vibration measurements is the amount of noise that the signal contains and the lack of knowledge available regarding the sources of the noise. Up to now, the prediction of failures has been based on analyzing the dynamic response of a machine to the excitation forces; meanwhile, the problem associated with noise is that the CMS relates failures to the dynamic response at specific frequencies and their variation along the life of the equipment. However, the noisy response may contain information regarding early failures that requires another type of signal analysis besides the most utilized signal analysis, which is the Fourier transform.

CMS require sophisticated algorithms to process all the signals. The machine is outfitted with a sufficient number of sensors that record data. The set of sensors measures vibrations, temperature, current at the power supply, oil quality, thermography, sound, acoustic emissions, and any other variable related to the process of each particular machine train. Some of the variables can be measured in line (continuously) and others have to be measured periodically, and the periodic records reinforce the analysis of the continuous data.

Oil analysis provides information regarding the wear of metallic parts, the presence of debris, and oil degradation. There are continuous oil transducers that provide information regarding small debris and oil degradation while large debris can be identified with magnetic detectors that provide overall information that is good for alerts. With an intelligent algorithm, the signal produced by a magnetic detector serves as an input for retrofitting the continuous vibration measurements. In this way, the vibration measurement identifies the component that is failing, whereas the magnetic detector will determine the reason for the failure. For example, if the analysis of the vibration signal indicates that a roller bearing has incremented the vibration amplitude, the magnetic detector will confirm whether there is debris production, and the combination of the two signals with the process parameters will improve the decision on the type of failure. In this case, the presence of debris on the magnetic detector will indicate that the roller bearing has a crack and it is losing material. Other techniques complement the vibration measurement, such as thermography (Fig. 8.2), acoustic emissions, endoscopy, etc.

The best way to improve the performance of a CMS is by combining all the possible signals into an expert system that analyzes the data, produces different probable outputs, and prevents future failures. Because oil sensors only detect a few parameters, a proper oil analysis requires the application of laboratory tests on samples that are acquired in batches. It is important that the CMS should be designed in a way that the laboratory results (moisture content, viscosity change, detergent degradation, etc.) are regularly included in the expert system, and the data are synchronized with the corresponding continuous measurements (Fig. 8.1).

It is crucial to measure the process parameters to determine the level of power demand that the machinery train is carrying out. These parameters can include current and voltage on electric motors, pressure and temperature in thermal machinery, exhaust production in combustion engines, and the actual rotating speed.

CMS is the primary input for every Industry 4.0 (a subset of the fourth industrial revolution) because it is based on cyber-physical systems, and CMS sensors and algorithms can deliver data about the production condition, the prediction of maintenance operations, and the estimation of life cycles for products and processes. Industry 4.0 incorporates all this information and provides scenarios to make decisions in global environments (Fig. 8.2).

## Signal analysis

Signal analysis is the basis for determining monitoring conditions. A full discussion of vibrations and vibration measurements was held in previous

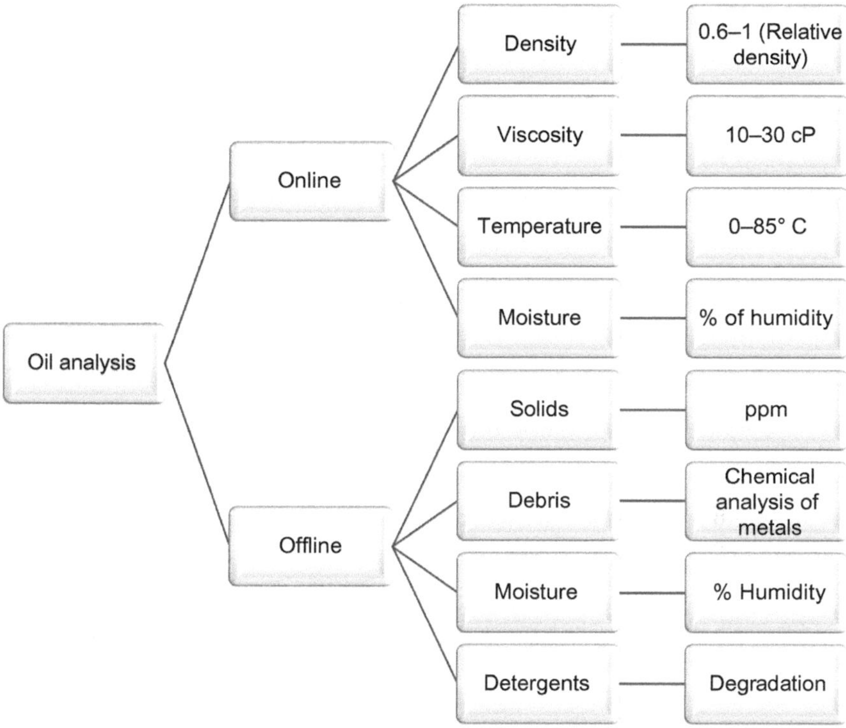


**Fig. 8.1** Structure of an oil analysis system.

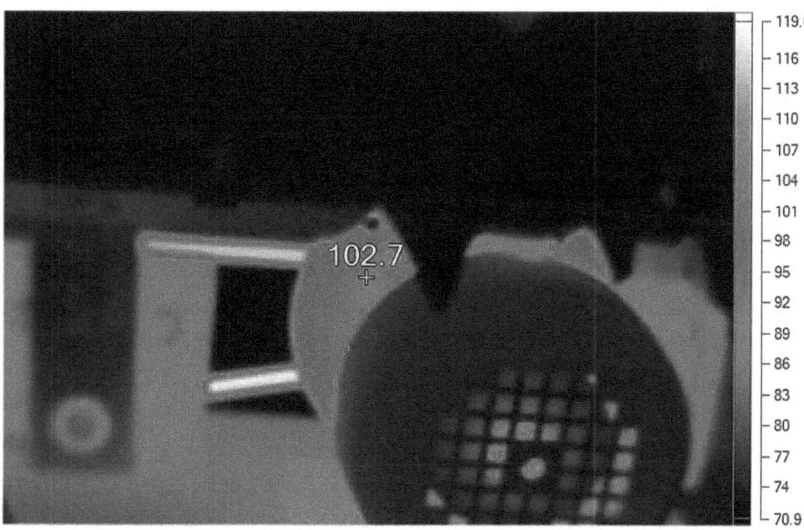


**Fig. 8.2** Typical picture of a thermography analysis.

chapters, and the type of sensors was also presented. It was also made clear that there are two main parameters to be considered: the range of frequencies and the amplitude of the vibration. For very low frequencies, it is recommended to use displacement or velocity sensors, whereas it is recommended to use accelerometers for higher frequencies. It is essential to consider the gain of the sensors because it determines the quality of the "shape" of the signal.

Because the behavior of a machine is time-dependent, the monitoring signal is also time-dependent, and it exposes the nature of the machine and helps understand its condition. The original data recorded with the sensors combine many frequencies and overlap the effects produced by every component. The data must be transformed to a different domain to overcome this difficulty. In previous chapters, the use of the Fourier transform was described. This mathematical procedure has been widely used, and it is well accepted by industry. In this chapter, other techniques are included, and their fundamentals are also described here. All these techniques for signal analysis are based on the convolution theorem.

The basis for signal analysis is the Fourier transform, which converts a data series x(t) into a series of functions in the frequency domain. It is based on the Laplace transform and is based on the concept that a periodic signal can be substituted by an infinite series of weighted sine and cosine functions. The Fourier transform has some limitations. It cannot identify the transient component of a signal, and when it is applied to nonlinear data, it produces frequency spectra that have no physical meaning.

To overcome the limitations of the Fourier transform, Dennis Gabor in 1946 presented a method that converts the signal data from a one-dimensional transformation into a multiple-dimensional solution. The technique is known as the short-time Fourier transform (STFT). His method is based on defining a "window" of a certain length, then sliding it along the signal. With this procedure, a time-localized Fourier transform is obtained, and the signal can be separated in blocks of similar frequencies that occur at different instants. The graphical representation of the blocks produces a two-dimensional plot that is known as the spectrogram of the time-frequency map (Fig. 8.3).

With this method, a signal can be segmented into a set of discrete Fourier transforms, and the signal is decomposed into a two-dimensional time-frequency graph or map. In a time-frequency map, or spectrogram, the frequencies are separated according to their incidence time, and the transient components and the nonlinear effects can be identified as part of the entire map. A linear system, with a single excitation frequency, will produce a

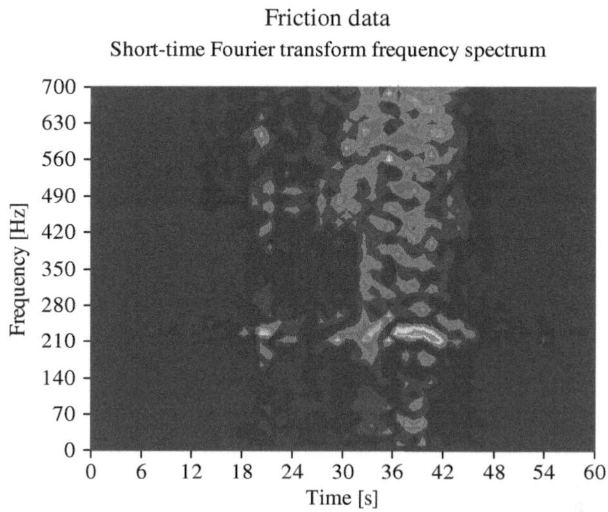


**Fig. 8.3** Short-time Fourier transform (friction data).

horizontal strip parallel to the time axle. Vibration measurements with different signal components that occur at different time intervals will show a pattern with many parts along with the entire map. Fig. 8.4 shows a sample of a spectrogram constructed using a signal; the data was taken from a rotor that rubbed the casing. The short-time Fourier transformation is obtained with

$$X(\omega, \tau) = \int_{-\infty}^{\infty} x(t)g(t-\tau)e^{-i\omega t}dt \tag{8.1}$$

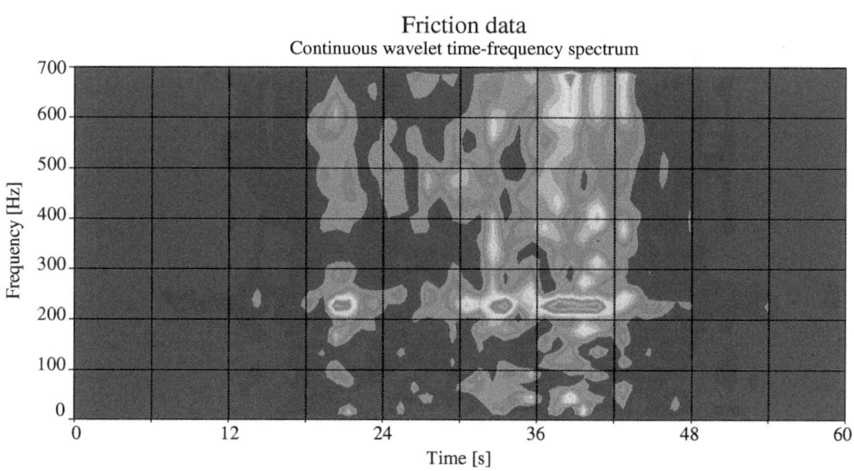


**Fig. 8.4** Spectrogram (continuous wavelet transform) of the friction data.

This equation can be interpreted as a measure of the similarity between $x(t)$ and the modulated window function $g(t)$. This segmentation produces a nonperiodic function, which creates discontinuities at the boundaries; thus, the Fourier transform creates more significant coefficients at high frequencies. For that reason, the windowing concept is introduced to reduce this effect. Instead of a rectangular segmentation, a smooth window function is employed to segment the original signal. The window function is almost one at the origin and zeroes at the edges. There are different functions for $g(t)$, as shown in Table 8.1.

A better method for producing spectrograms (time-frequency maps) is the continuous wavelet transform. The word wavelet comes from the French term "ondelette," which means small wave. Several researchers studied alternatives to Fourier analysis, but Jean Morlet was the first to implement a technique for scaling and shifting the analysis window function. He analyzed acoustic echoes produced by acoustic impulses sent to the ground and determined the existence of oil beneath the surface. His experiments marked the beginning of wavelets, as he and Alex Grossmann named them in the 1980s. Numerous researchers have followed their work. Stromberg did some early research on discrete wavelet transform, and Meyer and Mallat worked on the multiresolution analysis. Newman worked on harmonic wavelets for vibration analysis, and Ingrid

**Table 8.1** Window functions for the Short-Time Fourier Transform.

| **Function's name** | **Equation** |
|---|---|
| Barlett | $g(t) = a_0 - a_1\left\|\frac{t}{\tau} - \frac{1}{2}\right\| - a_2\cos\left(\frac{2\pi t}{\tau}\right)$ |
| Bisquare | $g(t) = (1 - t^2)^2$ |
| Gaussian | $g(t) = \frac{1}{\sqrt{2\pi}}e^{-\frac{t^2}{2}}$ |
| Hamming | $g(t) = \frac{25}{46} - \left(1 - \frac{25}{46}\right)\cos\left(\frac{2\pi t}{\tau}\right)$ |
| Hann | $g(t) = a_0 - (1 - a_0)\cos\left(\frac{2\pi t}{\tau}\right)$ |
| Kaiser-Bessel | $g(t) = \frac{J_a\left(\pi\alpha\sqrt{1-\left(\frac{2t}{\tau}\right)^2}\right)}{J_a(\pi\alpha)}$ <br> $J_\alpha$ is the zero-th order modified Bessel function |
| Welch | $g(t) = 1 - \left(\frac{t - \frac{\tau}{2}}{\frac{\tau}{2}}\right)^2$ <br> where $\tau$ is the vector length |

Daubechies proposed a family of discrete wavelets based on the concept of orthogonal multiresolution.

Wavelets have been applied to many different fields such as imaging processing, data condensation, solution of differential equations, etc. The wavelet transform decomposes a nonlinear signal into several frequency components. It preserves the time histories and will discriminate those frequencies that remain constant from those that vary along with the time domain. The regular frequencies are related to linear behavior, whereas the varying frequencies are related to the nonlinear terms.

The mathematical form of the wavelet transform is obtained from the convolution theorem:

$$X(s,\tau) = \frac{1}{\sqrt{s}} \int_{-\infty}^{\infty} x(t)\psi^*\left(\frac{t-\tau}{s}\right) dt \tag{8.2}$$

where $s > 0$ is the scaling parameter and it is inversely proportional to the frequency; it also determines the time and frequency scaling of the base function (namely mother wavelet). $\psi^*$ is the complex conjugated function of the mother wavelet.

For example, the Morlet mother wavelet is defined as

$$\psi\left(\frac{t-\tau}{s}\right) = e^{i2\pi f_0\left(\frac{t-\tau}{s}\right)} e^{-\alpha\frac{(t-\tau)^2}{s^2\beta^2}} \tag{8.3}$$

where $f_0$, $\alpha$, $\beta$ are constant.

Eq. (8.3) is formed by a sinusoidal function and a logarithm decrement function. This function is similar to the response of a second-order differential equation, and it can identify the similarities between the measured data for the reaction of a linear system. The time interval will be calculated as:

$$\Delta t = \frac{s\beta}{2\sqrt{\alpha}} \tag{8.4}$$

And the frequency interval as

$$\Delta f = \frac{\sqrt{\alpha}}{2\pi s\beta} \tag{8.5}$$

Small-scale values decompose high-frequency components, whereas large-scale values decompose low-frequency components.

The data from a rotor rubbing the casing was also analyzed with the continuous wavelet transform. Fig. 8.4 shows the spectrogram with this analysis.

## Identification techniques and condition indicators

CMS depends on signal analysis and many different techniques for analyzing the evolution of these analyses. The variations of the signals have to be done at specific periods. The periods are determined upon several parameters and the local changes of the data. Different parameters for evaluating these variations are presented in the following paragraphs.

### Intelligent systems

The procedure for identifying failures in machinery starts with the identification of the excitation frequencies (see previous chapters). These frequencies are created by all the components of the machine (The Annex presents a summary of these frequencies). Just identifying the frequencies associated with known failures limits the application of a CMS because other phenomena generate waveforms that differ from the traditional vibration signatures. That is the reason for including other monitoring devices and improving the signal analysis techniques.

Several condition indicators describe the evolution of failures in time. The application of these indicators requires the definition of two additional variables, the size of the sample (this value is determined from the sample rate and the frequencies of interest), and the period of sampling (which depends on the particular application, but a common criterion is to sample every 2 weeks if there is no significant change in the data). Some systems can be designed for continuous monitoring, but the size of the database needs to be massive; otherwise, it will be overloaded in a short time.

The signal analysis methods are designed for the analysis of the original data, and they produce either a frequency spectrum or a time-frequency map (spectrogram). These data are the samples that were referred to before, whereas the other indicators analyze the variations of the samples every other week.

### Time synchronous average

Time synchronous average is a method for reducing noise in complex signals, and it differs from spectral averaging. The waveform of the original signal is averaged in a time buffer before calculating the Fourier spectrum. The overall vibration level (RMS) and the time domain signal are synchronously averaged and filtered around the important frequencies (those produced by the mechanical components). The signal is sampled with a trigger (it could be a tachometer) that is synchronized with the original data. This process is repeated several times until the random noise is eliminated. The coherent signal will remain, and the higher noise frequencies are removed. The risk of using this procedure is that it can eliminate some signals produced by failures that are asynchronous and nonharmonic. Time synchronous averaging helps in the analysis of waveforms, particularly in gearboxes. Time synchronous averaging is a proper procedure for setting the baseline of a healthy condition because the incipient failures have not yet appeared, and the nonharmonic waveforms can be implemented. It is also useful for setting the reference plot in new techniques as the recurrence plots. This method is only valid for those failures related to synchronous vibrations; for the rest of the problems, it is recommended to keep nonsynchronous signals and the noise (especially for nonlinear problems).

With the resulting processed data, the condition indicators (CI) are defined, and the reference levels are established. These levels set the baseline for condition monitoring, and they are evaluated every time that a sample is recorded. The variations on the CI value show two features: a tendency and random modifications. When the tendency line crosses a threshold level, there is a high possibility of detecting a failure. If the slope of the tendency increases, the sample period must be reduced to reduce the likelihood of missing a significant change in the machine conditions (Fig. 8.5).

The CIs are defined from different perspectives. They can be set at every specific frequency and time interval when their analysis depends upon different factors. The most recommended CIs are described next.

## Variation on the RMS

The RMS of the data is evaluated every time a sample is recorded. The RMS is used as a parameter that measures the energy content of a signal. It can be applied as an indicator of the health status of a machine. This indicator establishes the tendency of the energy produced by the vibration. The big disadvantage of this parameter is that it vanishes the high-frequency, low-amplitude signals. The mathematical formula for the RMS is:

$$A_{RMS} = \sqrt{\frac{1}{N}\sum A_i} \tag{8.6}$$

where $A_i$ is the data vector containing the sample signal.

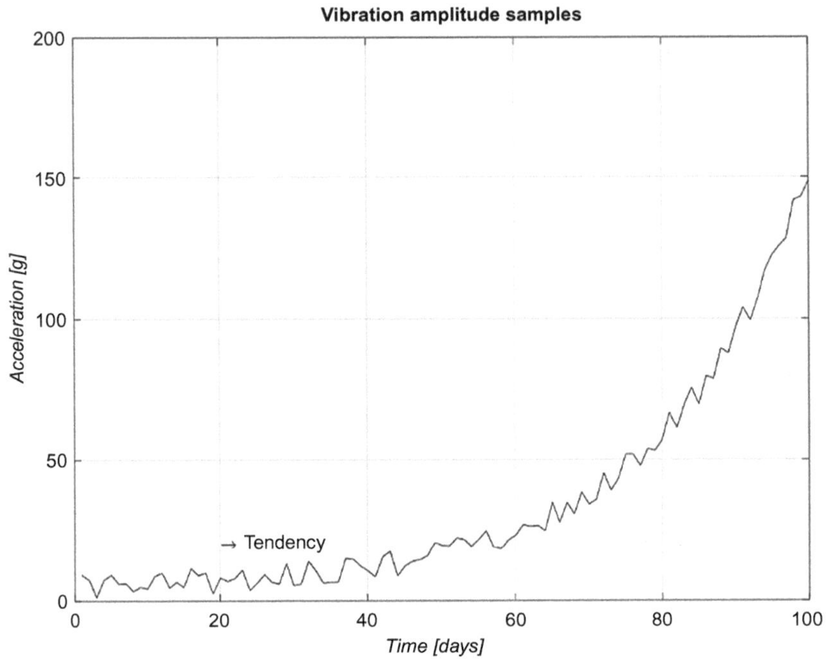


**Fig. 8.5** Evolution of the RMS factor during the sampling time.

## Highest peak

The tendency of the highest peak reflects changes in the machine condition and indicates whether there are significant variations in any of the elements of the machinery train. The disadvantage of this parameter is that the differences in the highest peak can be produced by operational changes and not necessarily by any failure or damage. It could provide false alarms. The highest peak can be evaluated at each of the significant frequencies (those defined in the Annex).

$$A_H = \max(A_i) \tag{8.7}$$

It complements the analysis of the RMS.

## Sideband level factor

The sideband index (or level factor) is defined as the average amplitude of the sidebands of the selected frequencies. This parameter measures the increment on the sideband amplitudes, and it is advantageous when the machine modulates signals, when the sidebands vary, or when one of the components presents transient or nonsteady operation conditions.

## Standard deviation

It calculates all the variations around the signal mean value. It gives similar information as the RMS value unless the signal has a biased form. For such cases, it is better to use the standard deviation instead of the RMS.

$$S_D = \sqrt{\frac{1}{(N-1)}\sum (A_i - \overline{A})} \tag{8.8}$$

In this formula, $\overline{A}$ is the average value and $N$ is the length of the data vector.

## Crest factor

This factor indicates variations in the signal amplitude related to the energy content. Any difference in a peak modifies the value, and it can be an indicator of a failure. A sinusoidal waveform with a factor $\sqrt{2}$ and a value around 6 is an indicator of damage.

$$CF = \frac{A_{i+1} - A_i}{A_{RMS}} \tag{8.9}$$

This CI amplifies local variations and distinguishes failures associated with high-frequency contents.

## Shape factor

It represents the standard deviation in the time domain of a signal divided by the overall signal value.

$$S_F = \frac{\sqrt{N\left[\sum A_i^2\right]}}{\sum |A_i|} \tag{8.10}$$

## Energy measurement

The energy measurement is the normalized kurtosis from the signal. At each point of the original signal, the indicator is determined from the squared difference of two adjacent points

$$E = \frac{N^2 \sum \left[(A_{i+1}^2 - A_i^2) - \overline{A}\right]^4}{\sum \left[((A_{i+1}^2 - A_i^2) - \overline{A})^2\right]^2} \tag{8.11}$$

## Kurtosis

The kurtosis is the fourth-order standardized moment of a given signal that gives a measurement of the significant peaks of a signal. The kurtosis value of a Gaussian noise signal is three, whereas a signal having variable peaks has higher values

$$K = \frac{N^2 \sum \left[A_i - \overline{A}\right]^4}{\sum \left[\left(A_i - \overline{A}\right)^2\right]^2} \tag{8.12}$$

## Energy ratio

$$E_R = \frac{\sigma_f}{\sigma_r} \tag{8.13}$$

where $\sigma_f$ is the standard deviation of the signal with failure, and $\sigma_r$ is the standard deviation of a signal recorded when the machine operates in a healthy condition (baseline signal). This factor measure increments of failures with respect to a reference value. It is also calculated as the ratio between the RMSs of a sample and the baseline.

## Zero-order figure of merit

This is a parameter that defines the ratio between the peak-to-peak amplitude divided by the energy content of the reference frequency, which can be used to detect changes in the time synchronous average. The difference of this parameter with respect to the crest factor is that it compares the changes in the peak values with the signal time average, and the other compares the peak values with the energy content.

$$FMO = \frac{A_{peak-peak}}{\sum A_i} \tag{8.14}$$

$A_{peak-peak}$ is the maximum peak-peak amplitude.

## Flatness FM4

This index determines whether the signal has peaks or is flat. It considers a set of points in the original signal, such that

$$F_l = \frac{N\sum (A_i - \overline{A})^4}{\left(\sum (A_i - \overline{A})^2\right)^2} \tag{8.15}$$

It can be constructed by generating a reference signal that contains the vibration data produced by the machine elements that characterized the equipment (see the Annex). This signal is subtracted from the original data and the remaining signal will contain only Gaussian noise when the equipment is healthy. If a defect appears, the remaining signal will contain other values instead of Gaussian noise.

A variation of this parameter is using only the envelope signal, withdrawing the higher-order frequencies.

## Residual flatness index NA4

This parameter overcomes some of the deficiencies of the flatness factor because it considers the residuals of the current data from the reference signal. It stores previous samples and generates residual signals from each stored data. The fourth statistical moment is compared with the average of the standard deviation of the stored samples.

$$NA4 = \frac{N\sum (A_i - \overline{A})^4}{\frac{1}{M}\sum \left(\sum (B_i - \overline{B})^2\right)^2} \tag{8.16}$$

where $B_i$ are the stored samples, $\overline{B}$ is the mean value of each stored sample, and $M$ is the number of stored samples. This indicator reflects the spread of a failure along the sampling period and it is sensible to the increment in the amplitude.

## Envelope factor NB4

The envelope factor is similar to the residual flatness index, but it considers the envelope of the signal. The parameter is set at every individual desired frequency (see Annex for reference to the individual signal). The signal is filtered around the desired frequencies, and afterward the Hilbert transform is applied to each remaining signal. The resulting parameter is evaluated using:

$$NB4 = \frac{N\sum(E_i - \overline{E})^4}{\frac{1}{M}\sum\left(\sum(B_i - \overline{B})^2\right)^2} \quad (8.17)$$

where

$$E_i = \sqrt{A_i^2 + \widetilde{A}^2}$$

And $\widetilde{A}$ is the Hilbert's transform of the filtered signal:

$$\widetilde{A} = \mathcal{H}(A_i) = \frac{1}{\pi}\int_{-\infty}^{\infty} A_i \frac{d\tau}{t-\tau}$$

This parameter magnifies those fluctuations that are caused by any failure in the machine component.

## Shannon entropy

This indicator describes the distribution of the energy in a signal. It is defined as the probability that a value of $A_i$ appears in the signal (the probability of $A_i$ appearing in $A$, where $A$ is the vector that contains the sample data). Mathematically, it is expressed as:

$$H_E = -\sum_{j=1}^{n} p_j \log p_j \quad (8.18)$$

where

$$p_j = P(A_i | A)$$

## Case study

The following case study shows the application of different indicators to vibration measurements taken from a rotor that rubs the casing. This case is an example of a typical failure, with high nonlinearities. The data were sampled at three different events; these events were identified while the fault was evolving. This case study shows the effectiveness of the indicators because the failure can be detected at a dominant frequency. The indicators display different results as the failure evolves. Some of them reduce the significance of the failure; meanwhile, others are consistent with the increment in the damage. Fig. 8.6 represents a sketch of a rotor rubbing the casing. At the first stage, the rotor was free to rotate without hitting the involved structure; once the deflection of the shaft increased, the rotor hits the casing, and

**Fig. 8.6** Sketch of a rotor rubbing the casing.

the friction and radial forces are created. The radial deformation grows and the friction force is incremented, producing higher vibrations than at low friction.

The vibration measurements were recorded at the supports of the rotor, and they corresponded to three conditions: without friction, low friction, and high friction.

These three stages were sampled at different times, and they reflect the increment on the failure. Fig. 8.7 shows the recorded data. Fig. 8.8 includes the frequency spectrum of the three samples. It is clear that the highest peak is at the rotation speed, but there is another peak that has a different frequency depending on the load condition: at the low frequency it appears at 116 Hz whereas at the high frequency, it is noticeable at 80 Hz.

The evolution of the failure was evaluated with indicators that take into account the entire sample. For this case study, only the RMS, the Kurtosis, the Shannon entropy, and the shape factor were considered. These indicators measure the condition of the sample data with different results. Fig. 8.9 shows the evaluation of the RMS from the three stages. It is clear that the highest value occurs at low friction, but it was expected to have the maximum value with the most severe condition. The same situation is found with the Kurtosis (Fig. 8.10) and with the Shape factor (Fig. 8.11). The difference between these indicators is the sensitivity of the calculated values. The ratio between the highest value versus the lowest value is 1.7 for the RMS indicator, 2.5 for the Kurtosis, and 3.8 for the shape factor. Analyzing these results, it is clear that the shape factor is more sensitive to small variations in the sample data. The Shannon entropy (Fig. 8.12) provides different

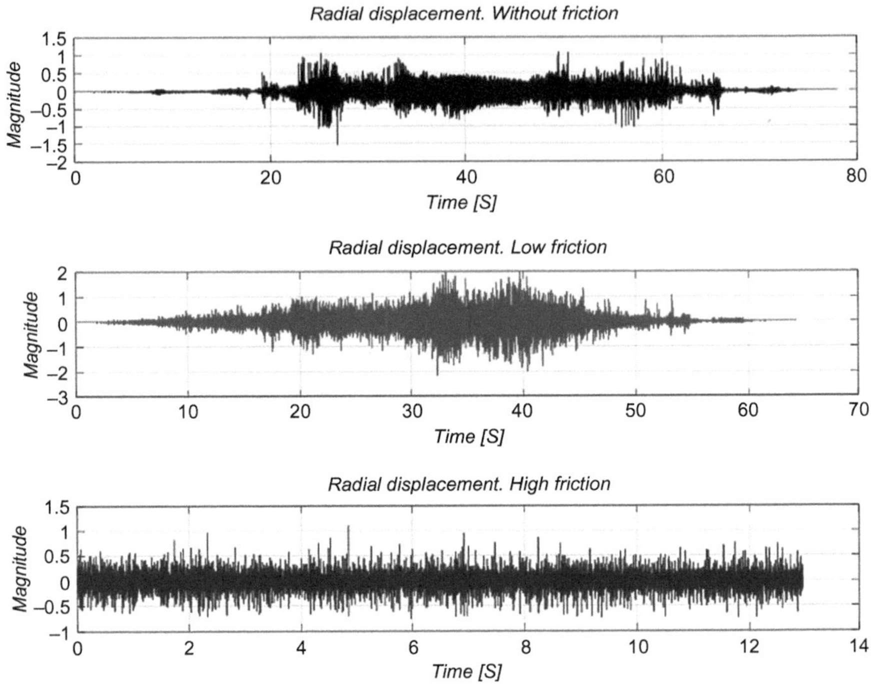


**Fig. 8.7** Original data.

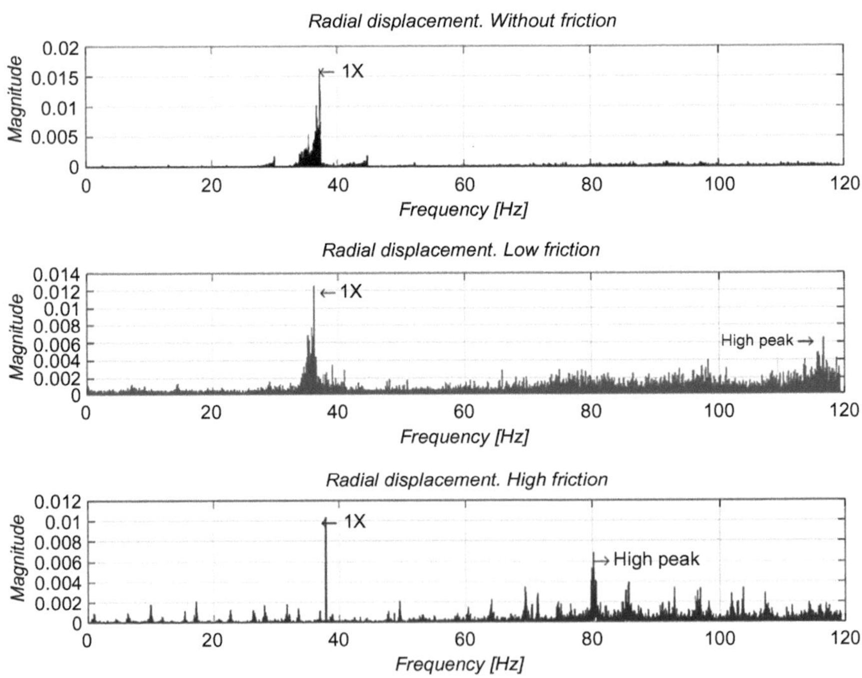


**Fig. 8.8** Frequency spectrum of three samples.

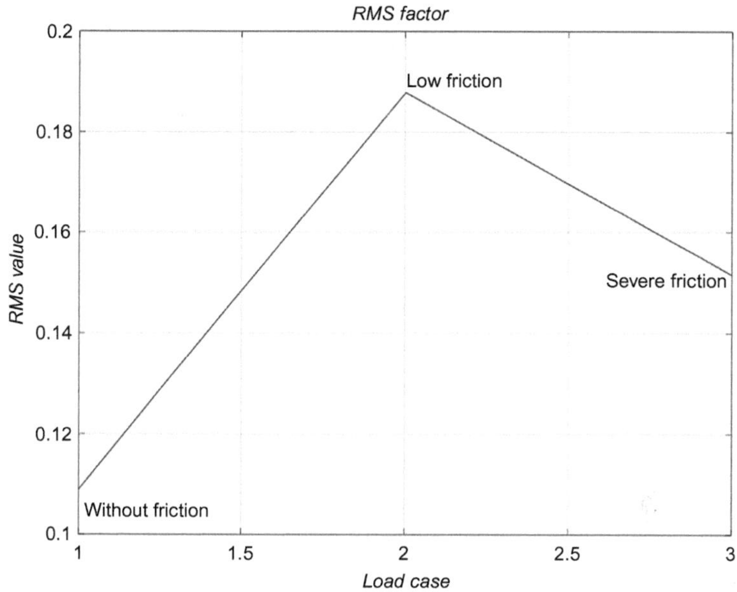


**Fig. 8.9** Evolution of the RMS indicator.

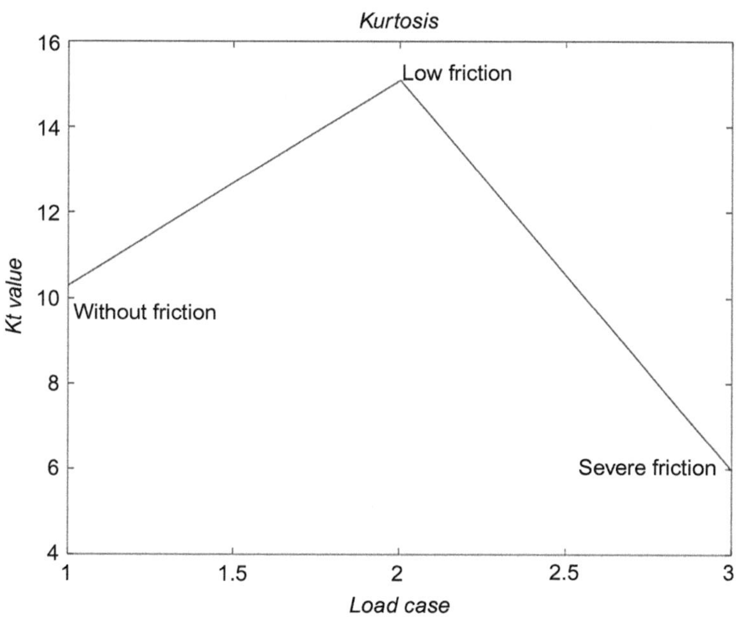


**Fig. 8.10** Evolution of the Kurtosis Indicator.

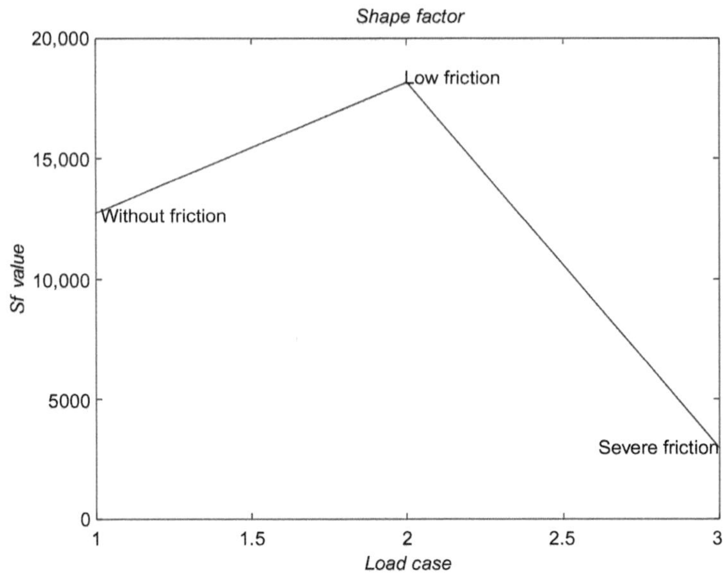


**Fig. 8.11** Evolution of the shape factor Indicator.

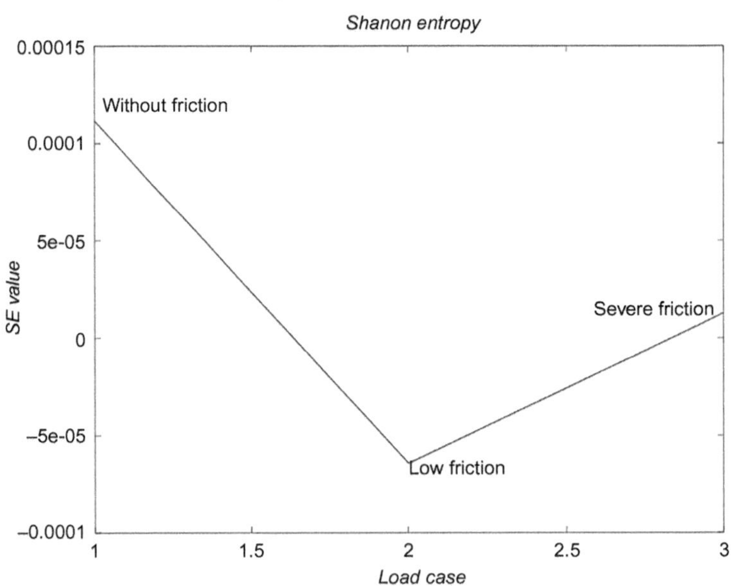


**Fig. 8.12** Evolution of the Shannon entropy.

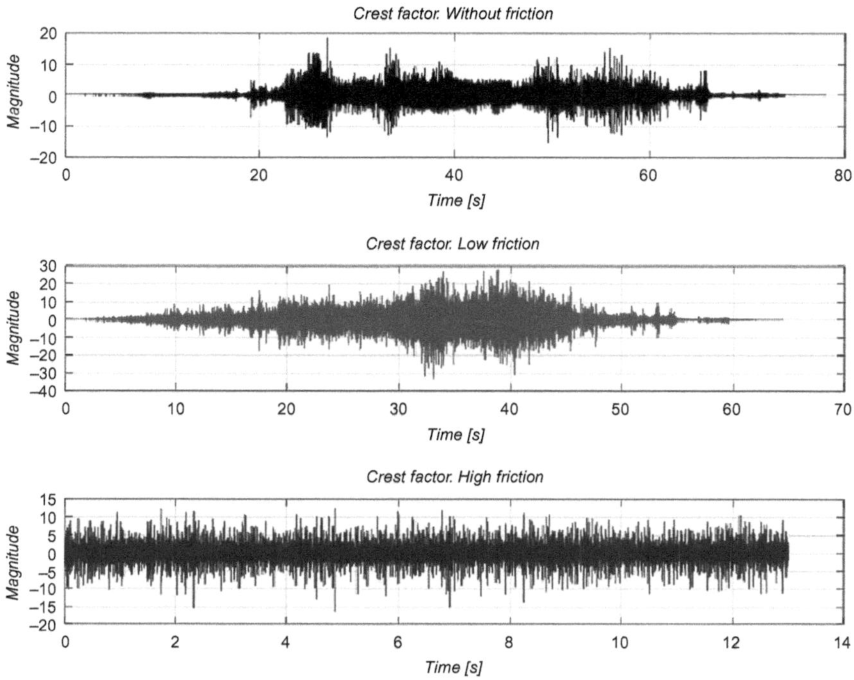


**Fig. 8.13** Crest factor indicator.

results in comparison with the other indicators. For this data, the lowest value corresponds to the "low friction" data and the highest value to the healthy state ("without friction"). Even though the tendency is opposite to the other indicators, it shows a similar relationship between the three stages.

Other indicators evaluate variations within the signal data. The measured data were analyzed with the crest factor (Fig. 8.13). These results are difficult to interpret, and it is necessary to process them with the FFT. Fig. 8.14 shows the frequency spectrum of the crest factor for the three conditions. When these spectra are compared with the original data (Fig. 8.8), it is found that the crest factor amplifies the high-frequency content of the signal and magnifies the evolution of the failure. This indicator gives a better evaluation of the development of the fault because it detects the "high frequency" state as the highest value at high frequencies.

The definition of the indicators requires analysis of the type of failures that will be monitored. The system should include overall indicators as well as global indicators. The crest factor and the Shannon entropy show better

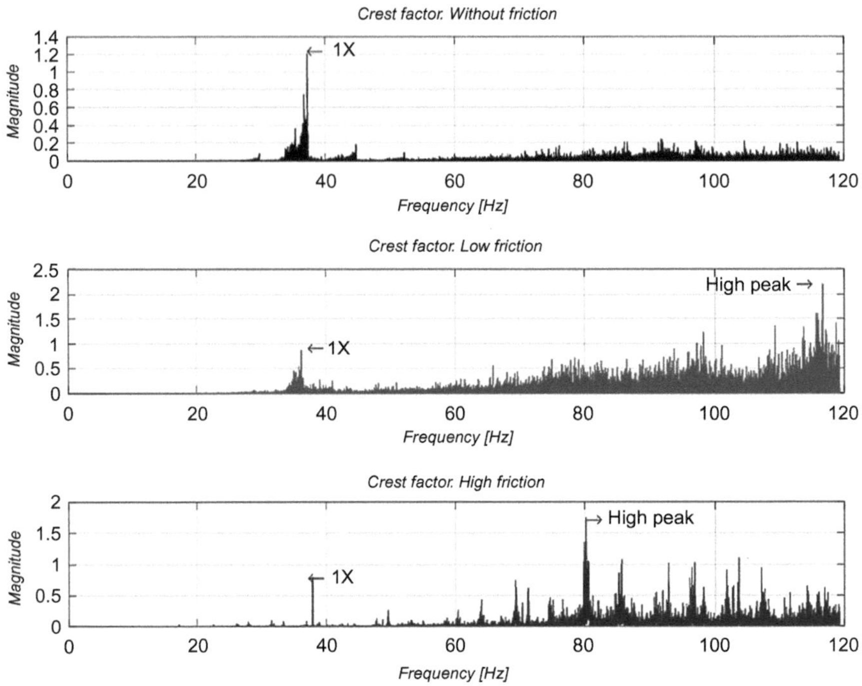


**Fig. 8.14** Frequency spectrum of the crest factor indicator.

results for the case study. This is because low friction is less chaotic and has better-defined peaks in the frequency spectrum, whereas the frequency spectrum of the "high friction" is more "noisy." The indicators must be selected according to the elements that will be monitored. They must have a combination of them to detect the tendency of each significant frequency (see the Annex for reference).

# Annexure

This chapter includes a summary of the most relevant failures and the typical spectrum. All spectra were created using simulated signals to emphasize the relevant response.

The analysis of the following failures is based on the assumption that two measurements are recorded simultaneously at both supports, and the phase angle between the two signals is measured in the time domain.

The first group of failures is related to defects in the rotor.

## Unbalance

**1.** Static unbalance

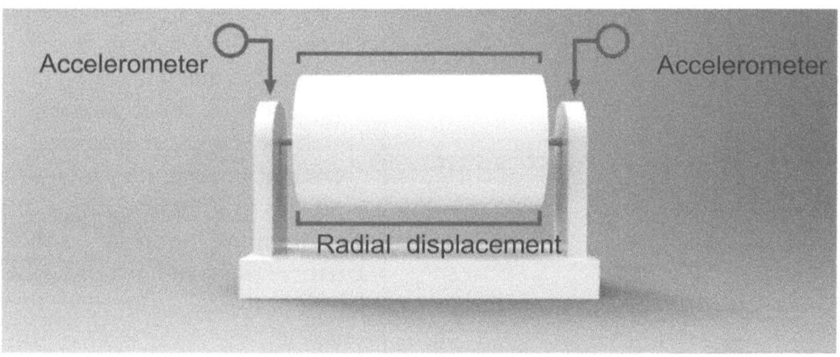


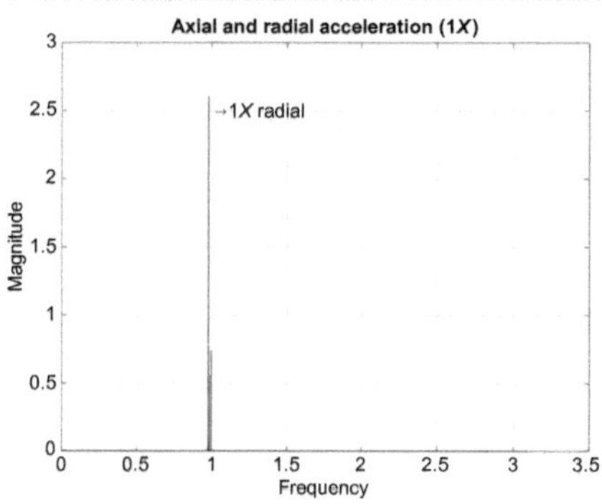


The static unbalance is also known as the unbalance in one plane. It occurs when the center of mass is located away from the rotating axis, the reacting force is divided equally between the adjacent supports, and it produces radial accelerations at the rotating frequency. The frequency spectrum displays a dominant peak at 1*X*, and the phase angle between the two supports is almost zero. The amplitude increases at the square of the speed increment. The rotor has to be balanced to reduce the vibration effects.

2. Dynamic unbalance

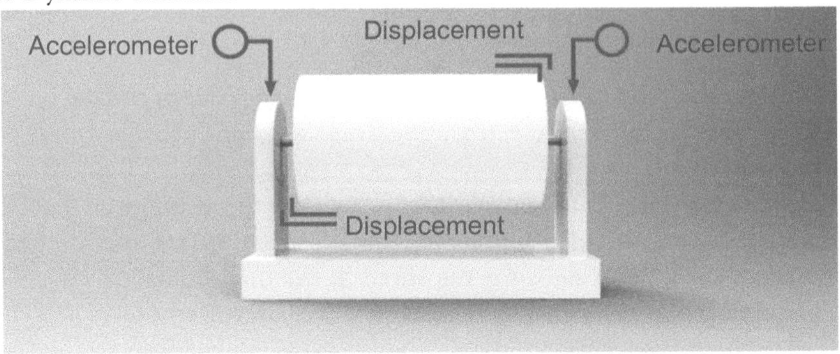


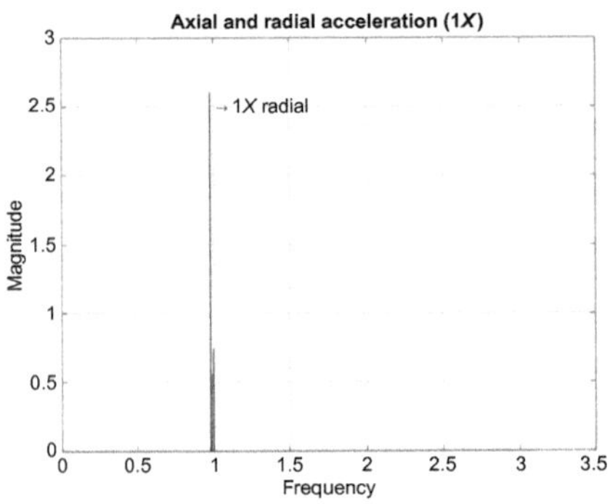


The dynamic unbalance is also known as a two-plane unbalance. In this case, the center of the mass is located away from the rotating axis, and in the axial direction, it is closer to one of the supports; therefore, the reactions at the bearings are uneven. The phase angle between the signals is 180 degrees, and the spectrum has a dominant peak at 1*X*, both in the axial and in the radial directions. The amplitude increases at the square of the speed increment. This problem can be reduced by balancing the rotor in two planes.

**3.** Unbalance of cantilever rotors

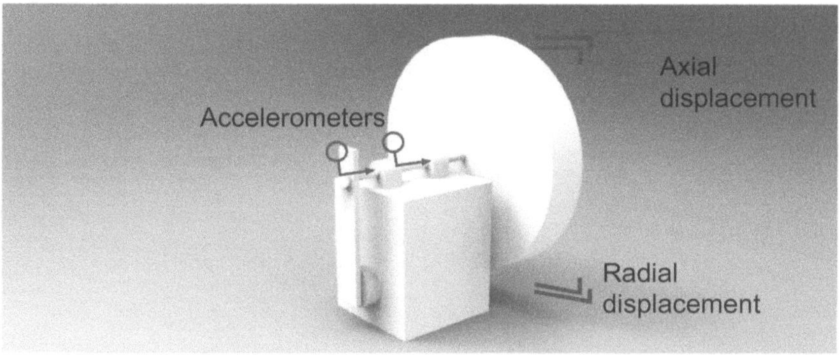


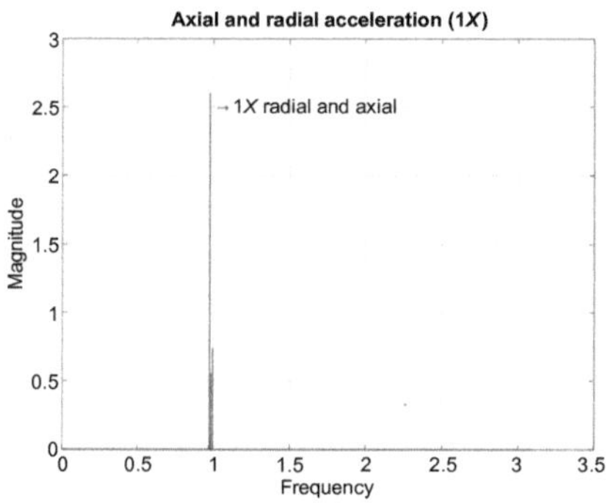


The rotors mounted in the cantilever show the static and dynamic unbalance. The main characteristic of this problem is that the dominant frequency is higher in the axial direction. The phase angle between the radial measurements is almost zero. The problem can be reduced by balancing the rotor in two planes, ideally one at the rotor and the other at the opposite extreme.

**4.** Eccentric rotor

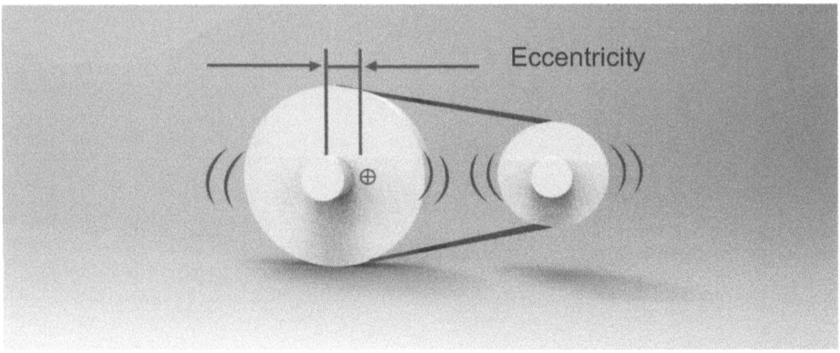


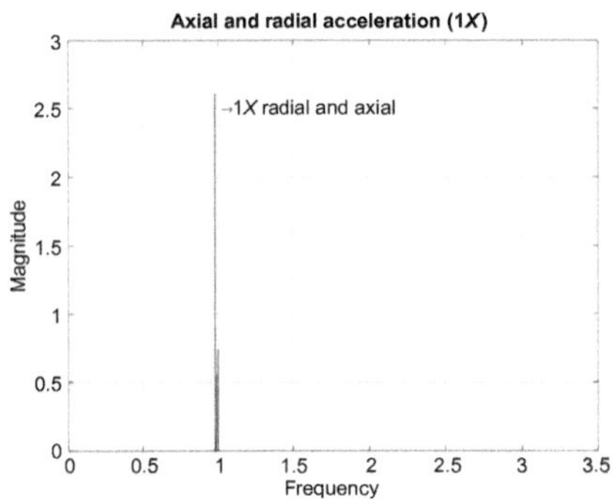


The rotors are eccentric when the center of rotation is offset from the geometric center. The dynamic response is similar to an unbalanced rotor, the dominant peak occurs at $1X$ in the radial direction, and the phase angle between the horizontal and vertical directions is 180 degrees. It is possible to reduce the vibration effect by balancing the rotor, but it is possible to reduce the vibrations in one direction only.

**5.** Bent rotor

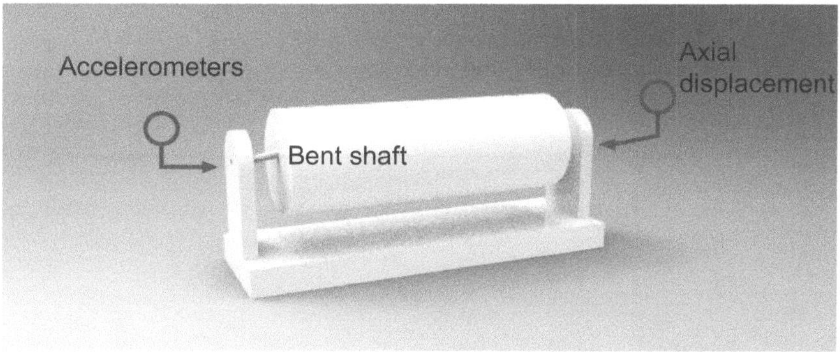


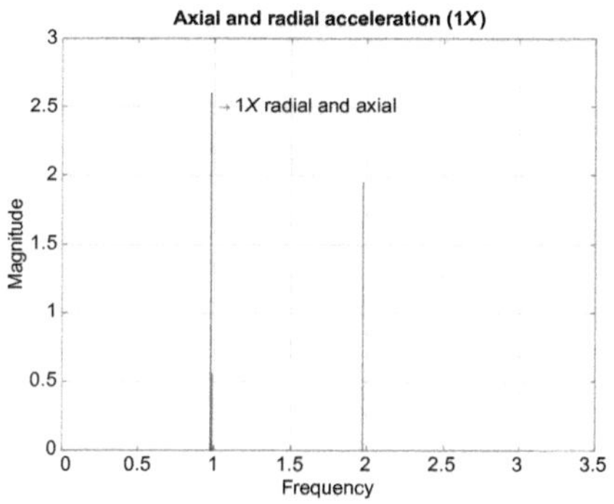


The shaft of a rotor can be bent for many reasons. The bend can be temporal or permanent, and the most common situation occurs when the shaft has a thermal gradient along its axis. The vibration spectrum shows a dominant peak at $1X$ and a second peak at $2X$. The phase angle between the support signals is 180 degrees.

## Misalignment

The second group of failures is related to misalignment problems.

**6.** Angular misalignment

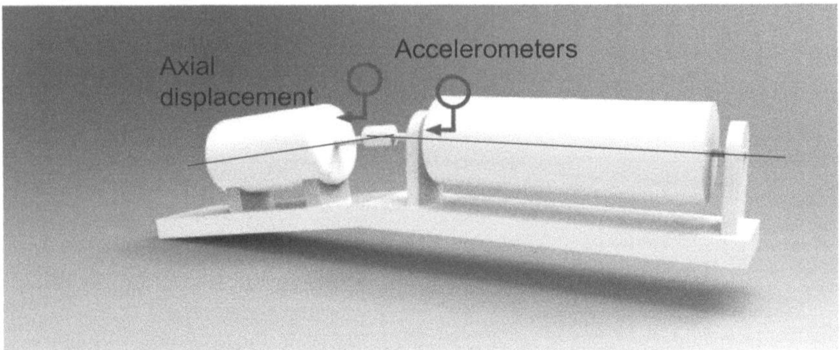


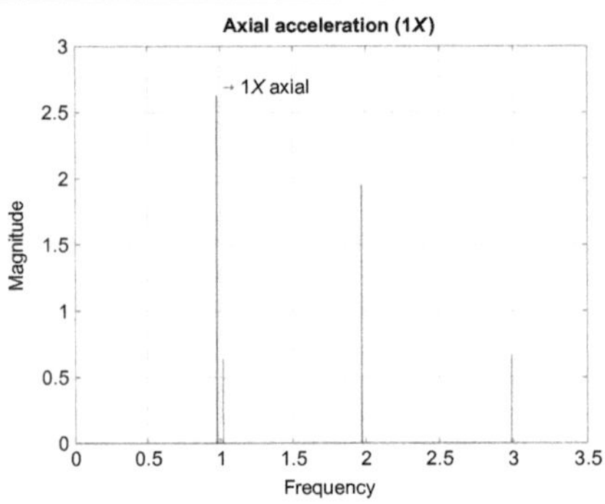


Misalignment is the most straightforward problem to solve, but it requires a systematic approach.

A high axial vibration characterizes the angular misalignment. The frequency spectrum shows high peaks at 2$X$ and 3$X$ in the axial direction. The amplitude is higher in the axial direction than in the radial direction. If the problem is sustained after aligning the equipment, it is necessary to look for other sources in the coupling.

**7.** Parallel misalignment

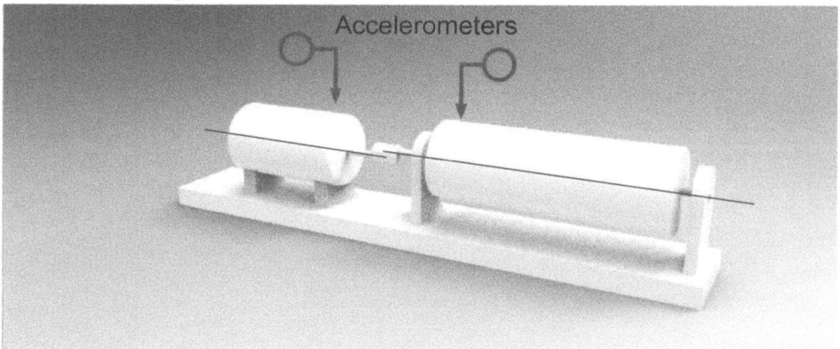


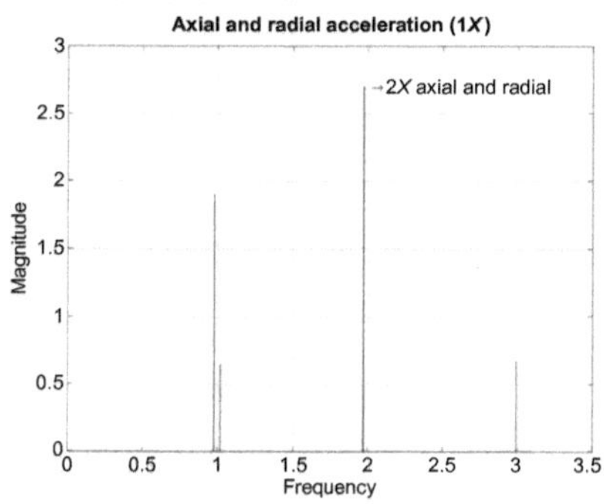


Parallel misalignment presents high vibrations in the radial direction. The axles orbit around each other and produce a high peak every half turn. The frequency spectrum shows a higher peak at $2X$ and another peak at $3X$. The phase angle between the supports at each side of the couple is 180 degrees. The severity of the vibration will depend on the type of coupling. Parallel misalignment is also a straightforward problem to solve.

**8.** Support misalignment

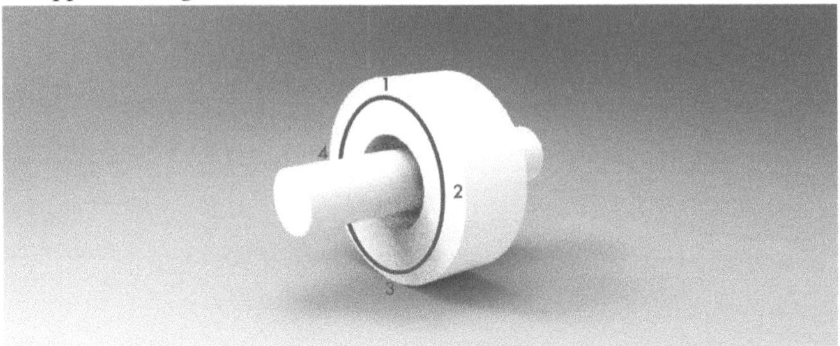


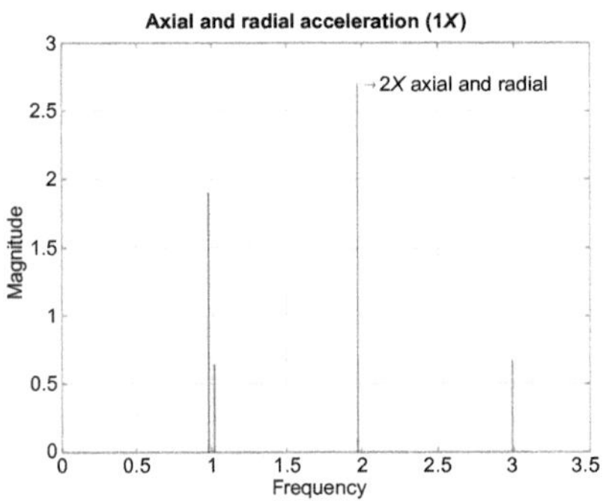


The support misalignment occurs when the center of rotation of both supports is offset. The identification of this problem could be cumbersome; it differs from other misalignment problems because the signal vibration at points 1 and 3 or 2 and 4 is 180 degrees out of phase.
The problem is solved when the support is replaced or adequately aligned.

## Transmissions

The third group consists of failures that appear at machine components in gearboxes and belt transmissions.

**9.** Gears

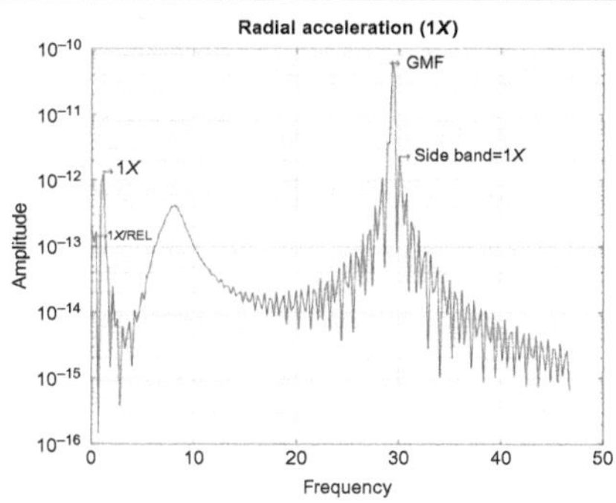


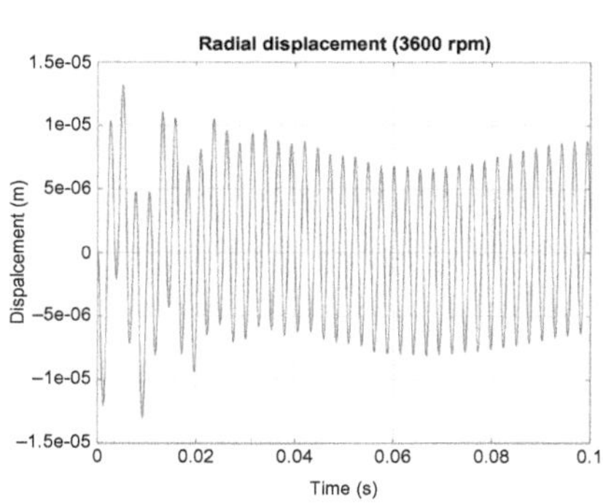


When the gear teeth are healthy, the spectrum shows two dominant peaks at 1*X* (pinion speed) and 1*X*/rel (at gear speed), a peak at the gear mesh frequency, and, in some cases, sidebands of low amplitude.

The wear on the teeth surface modifies the involute shape, and the motion is no longer smooth. This failure produces a spectrum with a higher peak at the gear mesh frequency (Np *X*) and one or two harmonics.

When the gear is eccentric, or the failure is located at several teeth, the time function shows a modulated signal and the spectrum. The distance between the sidebands corresponds to the frequency of the damaged gear. If both gears are damaged, the sidebands are double.

The presence of sidebands around the gear mesh frequency could be healthy. The frequency spectrum of a damaged gear shows sidebands at high amplitudes (almost at the same magnitude as the mesh frequency) and they are unsymmetrical.

**10.** Loose or worn belts

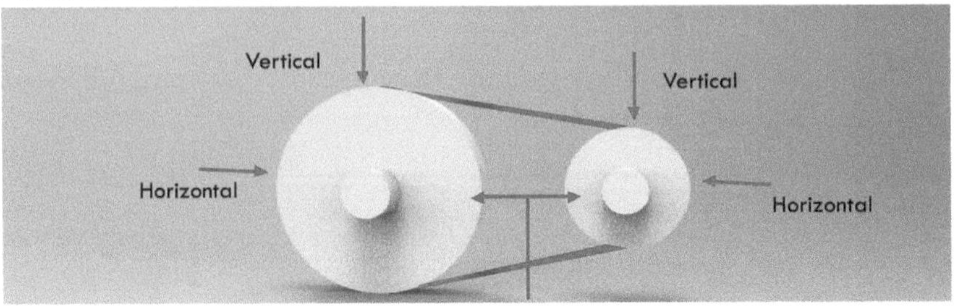


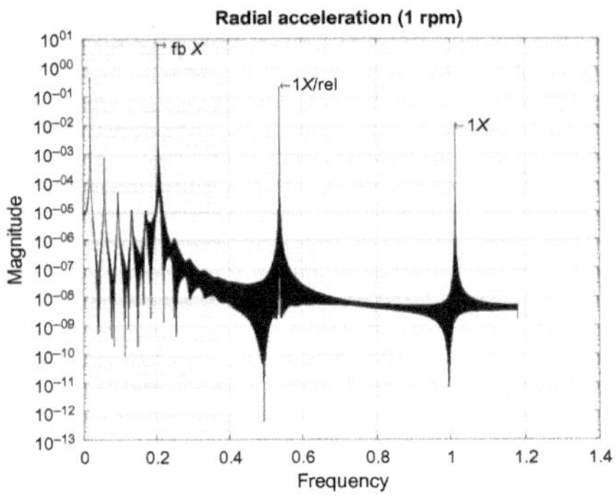


The belt transmission shows three characteristic peaks, one at the input pulley rotation speed (1*X*), a second at the output pulley speed (1*X*/rel), and a third at the belt flexural speed (fb *X*)

$$fb = \frac{input\ pulley\ pitch\ diameter}{belt\ length}$$

When the belt is loose or worn out, the frequency spectrum shows harmonics of the belt frequency (1*X*, 2*X* and 3*X*). The amplitude of the main peak is unstable

**11.** Misalignment in belts

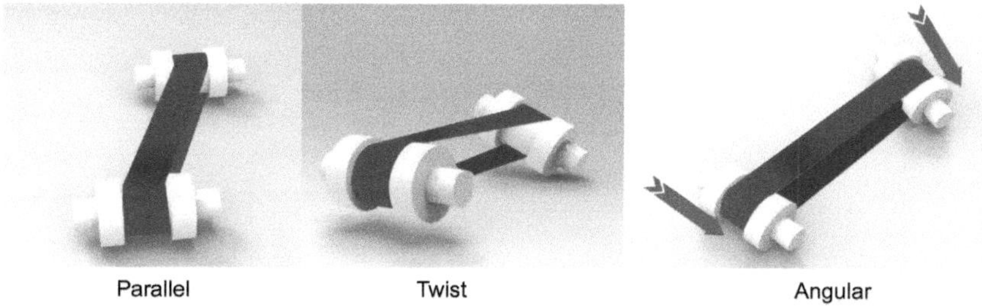


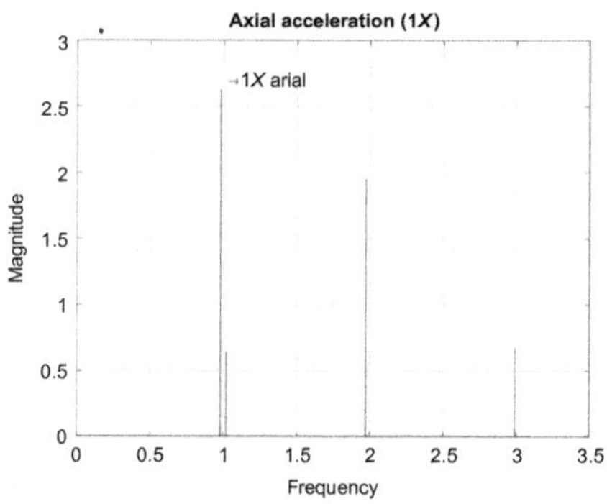


The misalignment produces higher peaks at the angular speed of the pulley where the sensor is mounted. If the sensor is located at the input pulley, the higher peak will be at 1*X*. If it is installed at the output pulley, the higher peak will be at 1*X*/rel.

The higher peak appears in both the radial and axial directions.

**12.** Resonance in belts

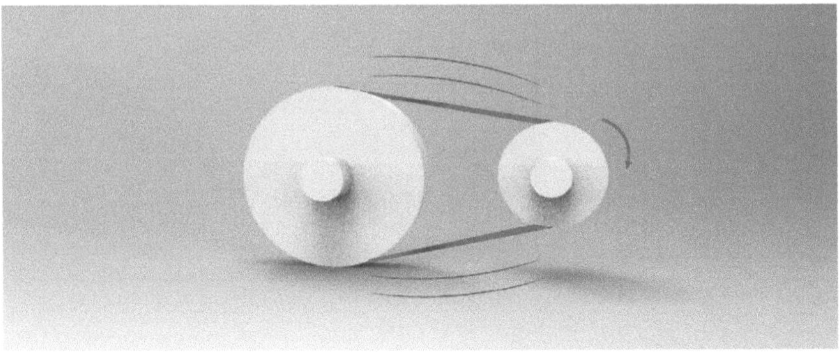

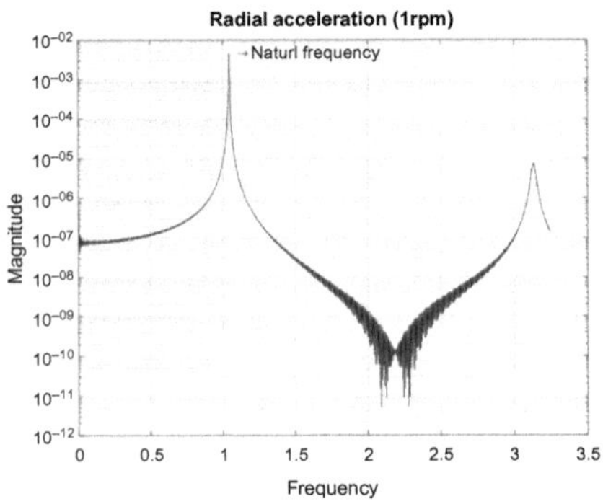


Belts have a particular condition because their natural frequency depends on the belt tension. When the rotating speed of either shaft is near the natural frequency, the belt will generate high noise and vibration. The most straightforward procedure to identify this failure is modifying the rotating speed of the input shaft until the noise reduces significantly. If the equipment cannot change the speed, then the fault can be reduced by modifying the tension on the belt. The spectrum will show a higher peak at the natural frequency, and this could coincide with any of the excitation frequencies described before.

**13.** Roller bearings

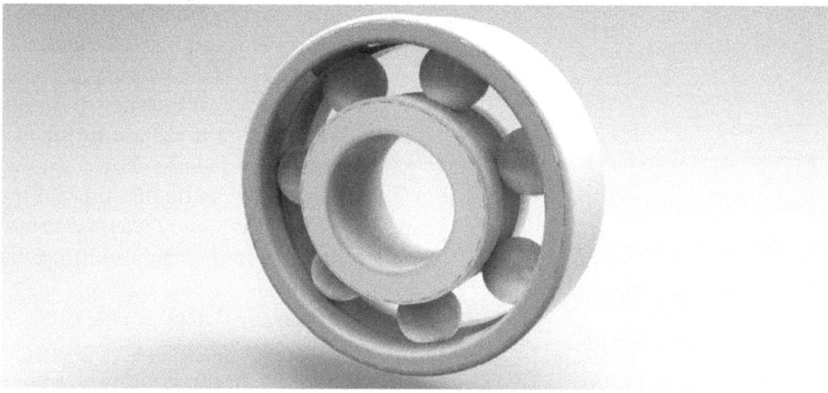

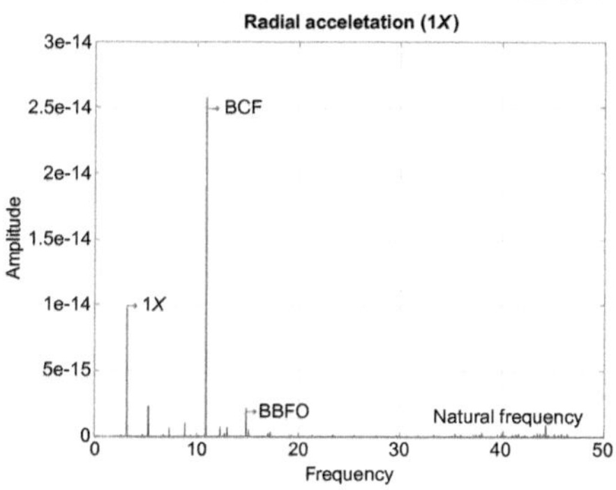


Bearing failures are the most critical faults in any machine. The bearing kinematics produce four typical excitation frequencies:

- Contact frequency between the roller element and the internal track

$$BBFI = \frac{N}{2}\left[1 + \frac{d}{D}\cos(\alpha)\right]X$$

- Contact frequency between the roller element and the external track

$$BBFO = \frac{N}{2}\left[1 - \frac{d}{D}\cos(\alpha)\right]X$$

- The casing frequency

$$BCF = \frac{1}{2}\left[1 - \frac{d}{D}\cos(\alpha)\right]X$$

- The roller spin frequency

$$BRF = \frac{D}{d}\left[1 - \left(\frac{d}{D}\cos(\alpha)\right)^2\right]\omega$$

When the bearing is healthy, the four particular frequencies have very low amplitudes, and only very high frequencies are present.

| | |
|---|---|
| | When a defect appears in one of the bearing elements (casing, roller, or tracks), the natural frequency of the bearing becomes evident. The frequency spectrum shows more peaks around the natural frequency. This condition can be identified if the natural frequency was estimated previously (before mounting the bearing).<br>When the defect grows, the particular frequencies become evident. At this condition, one or more peaks are present, and the spectrum can display one or two harmonics. It is essential to change the bearing when this condition appears. When the failure increases, more harmonics appear in the frequency spectrum.<br>At the end of life, the frequency spectrum will show a higher peak at the rotating speed (1*X*) and "noise" in a broad band of frequencies. This response is the result of the nonlinear behavior of the bearing. |

**14.** Journal bearings

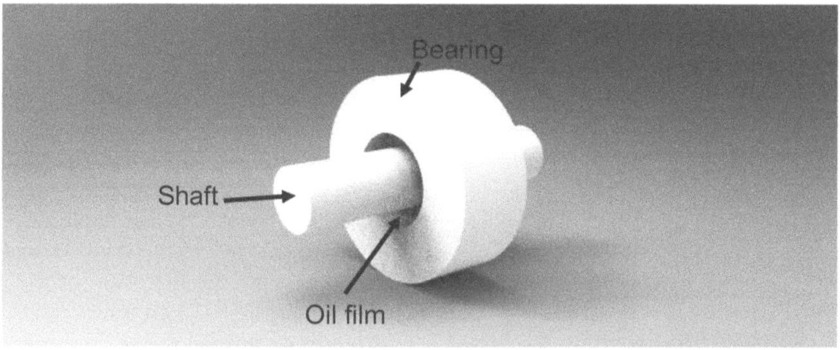


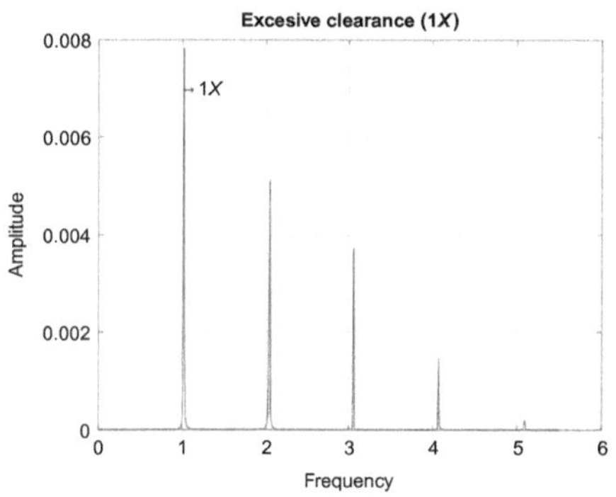


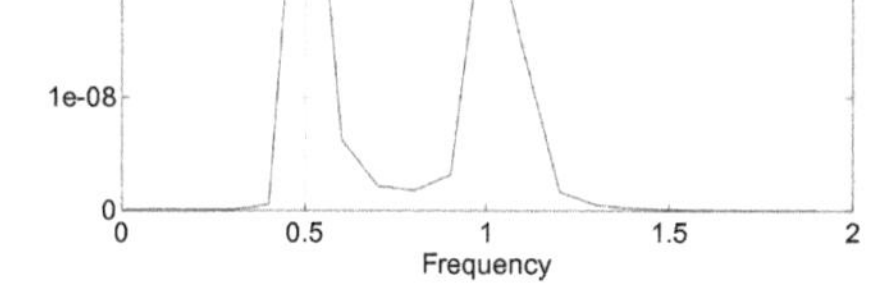


Journal bearings have a critical behavior that has led to much research and many publications. They have different designs but in general, they are based on the hydrodynamic response of an oil film. In contrast with the roller bearings, the dynamic behavior of the journal bearing is nonlinear, and the stiffness and damping depend upon the oil properties and the angular speed of the shaft. They present different failures that can be grouped in bearing wear (excessive clearance), oil whirl, oil whip, and rub. There are other problems such as extreme heat, oil degradation, and bearing fractures.

Failures that are related to an ample clearance between the axle and the bearing and it appears as an unbalance with multiple harmonics of the fundamental speed (up to 10*X*), and the vertical displacements could be larger than the horizontal movements.When the operating conditions have a sudden change, the oil film becomes unstable and creates an oil whirl. The

| | |
|---|---|
| 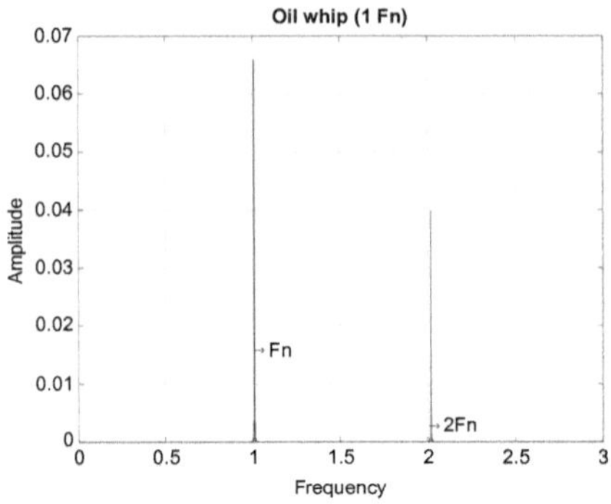 | operation conditions that cause this failure are a sudden change in the magnitude and direction of the radial load. Therefore, the hydrodynamic response becomes unstable, and the center of the shaft bounces along the direction of the radial component of the load. This movement creates a whirl that forces the shaft to orbit in the opposite direction of the rotating speed. The whirl can reduce the thickness of the oil film, and the shaft can rub the bearing. The frequency spectrum will show subharmonics with large amplitudes in a range between $0.42X$ and $0.48X$.<br>The oil whip appears in machines with slender shafts that rotate close to two times the first natural frequency (first mode).<br>This condition creates a critical state similar to a resonant vibration, increasing the radial displacement of the shaft. At this point, the hydrodynamic response of the oil film becomes unstable. The effect will synchronize the radial vibration with the second natural frequency, and the frequency spectrum will "lock" a peak at this state ($2F_n$). |

**15.** Aero and hydrodynamic forces

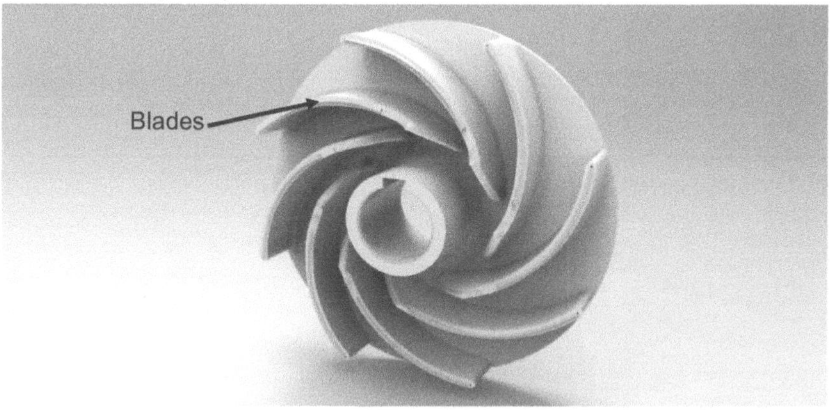


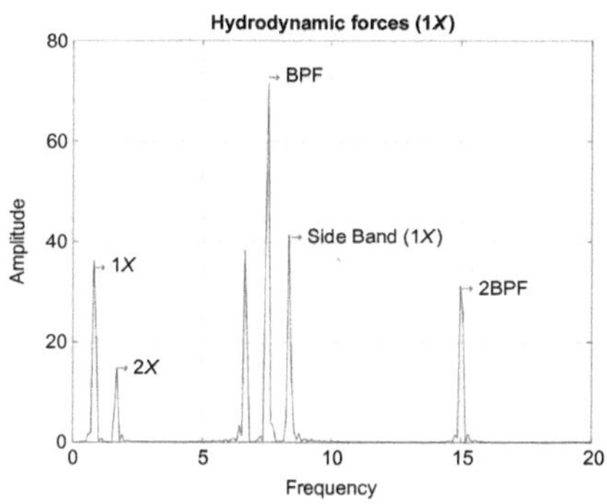


Hydraulic and pneumatic rotors show a dominant peak at the blade pass frequency (BPF). This frequency is the product of the rotating speed times the number of blades in the rotor. In a normal operation, the frequency spectrum will show two dominant peaks only, the rotating speed (1*X*) and the BPF. The BPF amplitude will increase if there is an excessive gap between the rotor and the casing. The frequency spectrum will show the harmonics of the rotating speed and BPF when the rotor is eccentric or misaligned.

The other failures that can be detected in the frequency spectrum are those related to the fluid. If the hydraulic flow is suddenly changed, or a drastic pressure drop appears in the pipeline, the flow could be turbulent. In this condition, the spectrum will show a noisy spectrum with a range of 0.8–33 Hz. If cavitation appears in hydraulic equipment, the spectrum will have very high amplitudes at BPF and a noisy spectrum at higher frequencies.

## Electric systems

The fourth group includes failures that appear in electric motors.

**16.** Eccentric rotor

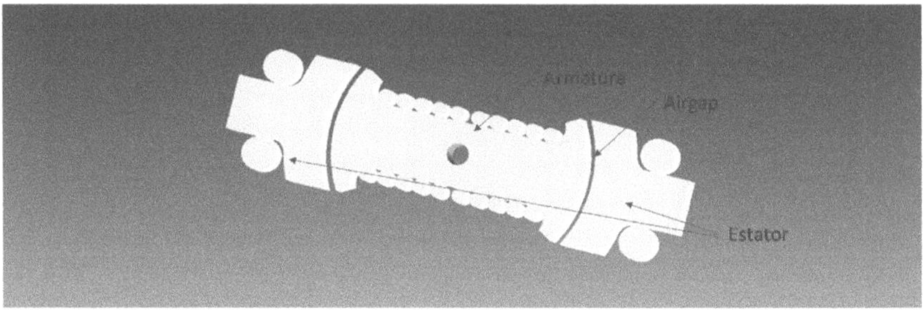


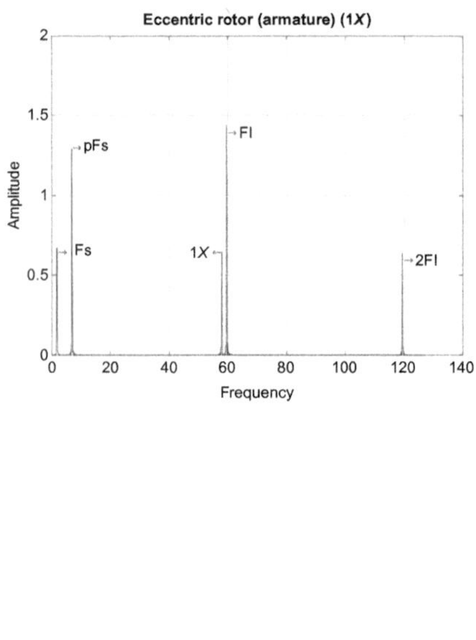


The frequency spectrum of an electric motor is characterized by these frequencies:

- Electric line frequency $F_l$ (50 or 60 Hz)
- Rotating speed ($1X$)
- Synchronous frequency $\left(\frac{F_l}{p}\right)$
- Slip frequency $\left(F_s = 1X - \frac{F_l}{p}\right)$
- Pole pass frequency $pF_s$

Where $p$ is the number of poles. The eccentricity in the rotor causes a variable air gap with the stator. This air gap variation modifies the electric frequency or line frequency $F_l$ and shows a dominant peak at $2F_l$.

When the armature (rotor) windings are broken or loose, or the lamination is in shortcut, the frequency spectrum will show higher peaks around $1X$ with

| | |
|---|---|
| | sidebands at the pole pass frequency.<br>When the armature windings are loose, the frequency spectrum will display a peak at $N_w X$ (number of armature winding times the rotating speed); if the damage is severe, the spectrum will show peaks at its harmonics.<br>If an electric connector is loose, the $2F_l$ will show a dominant peak and it will remain present at any velocity.<br>One procedure for detecting electrical problems is disconnecting the motor and observing those peaks that disappear when there is no electrical current in the machine.<br>One procedure for identifying the electric failures is turning the electric supply off and letting the machine rotate freely. The peaks associated with the electric problems will disappear, and the remaining peaks will depend only on the mechanical failures. |

# Other measurable failures

**17.** Resonance

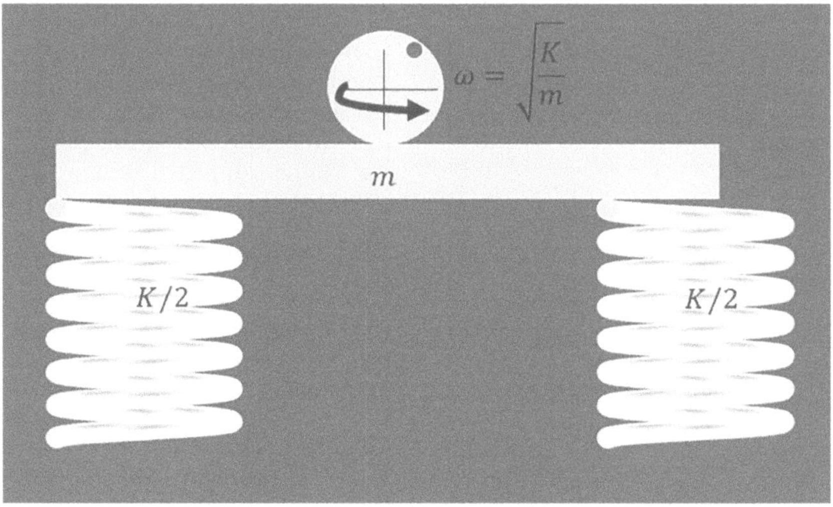


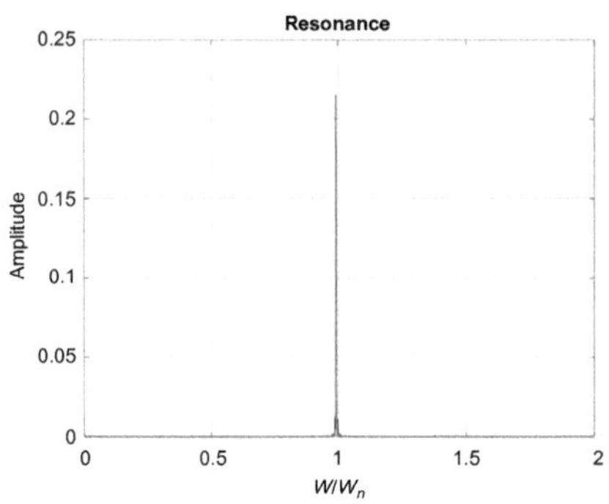


Resonance is a phenomenon that appears when the excitation frequency coincides with one of the natural frequencies of the machine or system. At this condition, the vibration amplitude severely increases, and it can cause a drastic failure. The vibration amplitude is inversely proportional to the damping characteristics of the system, which could be beneficial in some situations.

It is almost impossible to balance the rotor when it operates near one of its natural frequencies.

The identification of this failure can

| | |
|---|---|
| | be found by varying the operation speed. When the operation speed changes, the frequency spectrum will show two peaks (1*X* and the natural frequency). Because 1*X* depends on the rotational speed, its value will change, whereas the natural frequency will remain at the same value. Another way of finding this failure is through the analysis of the Campbell diagram. |

**18.** Looseness

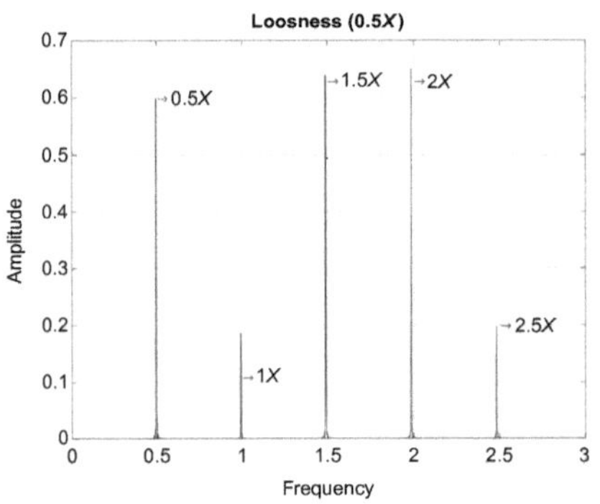


Looseness is a nonlinear problem that appears when one of the machine elements hits the structure. There are several locations of the hitting: when the machine is untightened at the foundation, when the ball bearings or the journal bearing housing is loose, or when one of the elements is improperly fixed to the axle.

In almost all cases, the frequency spectrum will show harmonics and subharmonics of the fundamental frequency (0.5$X$, 1$X$, 1.5$X$, 2$X$, …).

The failure is easily found when the machine is improperly mounted to the basement. The vibration measurements at the machine will be considerably more significant than the measurements at the foundation, and the measures at the vertical direction will be different from the horizontal ones.

The other failures require more analysis such as thermography to identify the location of the vibration source.

**19.** Rotor rubbing

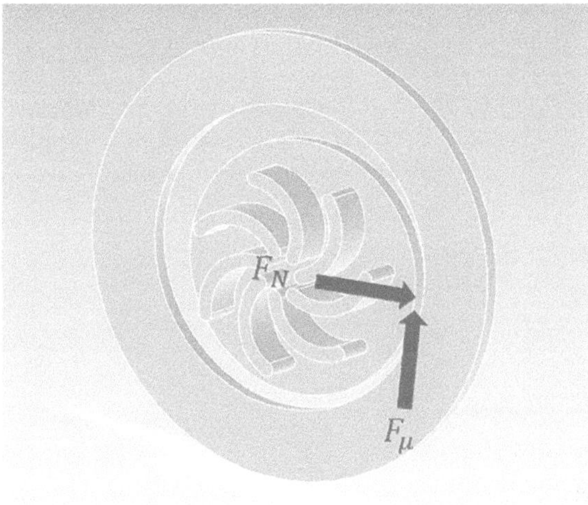


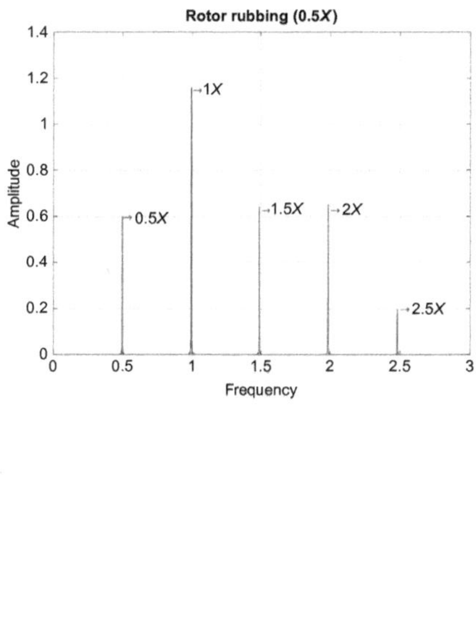


Rotor rubbing is a nonlinear phenomenon that appears when the rotor hits the casing. The frequency spectrum is similar to the one produced by a looseness failure; the difference is that the location can be identified easily. The rotor can rub the casing at a single point (local rubbing) or the entire circumference (complete rubbing). When the rubbing force is small, the frequency spectrum will show the harmonics and subharmonics of the fundamental frequency (0.5*X*, 1*X*, 1.5*X*, 2*X*, ...). When the rubbing increases, the spectrum will show a noisy response and the harmonics and subharmonics will disappear, displaying some unstable peaks at asynchronous frequencies. This condition can be challenging to identify, and it is recommended to use other measurements such as thermography.

# Index

Note: Page numbers followed by *f* indicate figures and *t* indicate tables.

## D



## E



## F



## G



## H



## I



## J



## K



## L



## M

## W



## Z

FSC
www.fsc.org
MIX
Papier | Fördert
gute Waldnutzung
FSC® C083411

Kolibri 360 GmbH
Ferdinand-Jühlke-Straße 7
99095 Erfurt, Deutschland
produktsicherheit@kolibri360.de